Lecture Notes in Mathematics

Edited by A. Dold and B. Eckmann

1310

Olav Arnfinn Laudal
Gerhard Pfister

Local Moduli and Singularities

Springer-Verlag
Berlin Heidelberg New York London Paris Tokyo

Authors

Olav Arnfinn Laudal
Universitetet i Oslo, Matematisk Institutt
P.O. Box 1053 Blindern, 0316 Oslo 3, Norway

Gerhard Pfister
Humboldt-Universität zu Berlin, Sektion Mathematik
1080 Berlin, Unter den Linden 6, German Democratic Republic

Mathematics Subject Classification (1980): 14D15, 14D20, 14B07, 14B12, 16A58, 32G11

ISBN 3-540-19235-2 Springer-Verlag Berlin Heidelberg New York
ISBN 0-387-19235-2 Springer-Verlag New York Berlin Heidelberg

Printing and binding: Druckhaus Beltz, Hemsbach/Bergstr.
2146/3140-543210

CONTENTS

NOTATIONS

k field

W^{*} the dual of the k-vektor space W

$k[\underline{x}] = k[x_1,\dots,x_n]$

$\underline{P}^n$ the projective n-space

$\underline{m}_S$ the maximal ideal of a local ring S

$S^{\wedge}$ the completion of the local ring S

$\underline{\ell}$ the category of local artinian k-algebras with residue field k

$\underline{\ell}_{H^{\wedge}}$ the category of local artinian $H^{\wedge}$-algebras with residue field k

$\underline{gr}$ the category of groups

$- \otimes - = - \otimes_k -$

$\underline{R} = \mathrm{Spec}\, R$, R a commutative ring

$\mathrm{Der}_k^c(A)$ the continuous derivations of a complete local k-algebra A

Def_X the deformation functor of X — §1

A^i the generalized André cohomology (see [La 1]) — §1,3

H^i the André cohomology for algebras (see [An],[La 1]) — §2

T^i $\mathrm{Sym}_k(A^{i*})^{\wedge}$ (see [La 1]) — §1

$\underline{H}^{\wedge}$ the formal moduli of X, i.e. the (prorepresenting) hull of Def_X — §1

$\underline{H}^{\wedge}_o$ the prorepresenting substratum of $\underline{H}^{\wedge}$ — §1

$\underline{H}^{\wedge}_{(n)}$ the n-th equicohomological substratum of $\underline{H}^{\wedge}$ — §1

$\pi^{\wedge}: \underline{X}^{\wedge} \to \underline{H}^{\wedge}$ the formal versal family — §1

$\mathrm{aut}_S(X\otimes S) = \{\phi \in \mathrm{Aut}_S(X\otimes S) \mid \phi\otimes_S k = 1\}$, S in $\underline{\ell}$ — §1

$\underline{\mathrm{aut}}_X$ the group-functor defined by $\underline{\mathrm{aut}}_X(S) = \mathrm{aut}_S(X\otimes S)$ — §1

$\mathrm{aut}_R(X^{\wedge}\otimes_{H^{\wedge}} R) = \{\phi \in \mathrm{Aut}_R(X^{\wedge}\otimes_{H^{\wedge}} R) \mid \phi\otimes_R k = 1\}$, R in $\underline{\ell}_{H^{\wedge}}$ — §1

$\underline{\mathrm{aut}}_{X^{\wedge}}$ the group-functor defined by $\underline{\mathrm{aut}}_{X^{\wedge}}(R) = \mathrm{aut}_R(X^{\wedge}\otimes_{H^{\wedge}} R)$ — §1

$\mathrm{aut}_R(H^{\wedge}\otimes R) = \{\phi \in \mathrm{Aut}_R(H^{\wedge}\otimes R) \mid \phi(\underline{m}_{H^{\wedge}}\otimes R) \subseteq \underline{m}_{H^{\wedge}}\otimes R\}$ — §2

$\underline{\mathrm{aut}}_{H^{\wedge}}$ the group-functor defined by $\underline{\mathrm{aut}}_{H^{\wedge}}(R) = \mathrm{aut}_R(H^{\wedge}\otimes R)$ — §1

$\underline{I}_{\pi^{\wedge}}$ the subgroup-functor of $\underline{\mathrm{aut}}_{H^{\wedge}}$ leaving $\pi^{\wedge}$ invariant — §2

$\underline{i}_{\pi}\wedge = \underline{i}_X$	the subgroup-functor of $\underline{I}_{\pi}\wedge$ inducing the identity on X	§2
$\ell(\pi^{\wedge})$	its Lie algebra	§2
$\pi:\tilde{X} \to \underline{H}$	an algebraization of the formal versal family	§3
$X(\underline{t}) = \pi^{-1}(\underline{t})$	for $\underline{t} \in \underline{H}$	§3
$\underline{H}_0$	the prorepresenting substratum of $\underline{H}$	§3
$\underline{X}_{\mu} \to \underline{H}_{\mu}$	the versal μ-constant family	§3
g_{π}: $\mathrm{Der}_k(H) \to A^1(H,\tilde{X},O_{\tilde{X}})$	the Kodaira-Spencer map of the family $\pi : \tilde{X} \to \underline{H}$	§3
$V=V(\pi), V_{\mu}$	the kernel of the Kodaira-Spencer map	§3,4
$\{\underline{S}_{\tau}\}$	the flattening stratification of the H-module $A^1(H,\tilde{X},O_{\tilde{X}})$	§3
$\{\underline{M}_{\tau}\}$	the local moduli suite	§3
$\mu m(f)$	the μ-modality of f	§4
$K(\underline{t})$	the matrix presenting $V_{\mu} \subseteq \mathrm{Der}_k(H_{\mu})$	§5
$L(\underline{t})$	the linear part of $K(\underline{t})$	§5

INTRODUCTION

The purpose of this monograph is to contribute towards a better understanding of the local moduli problem in algebraic geometry. Let k be a field, and let X be an algebraic object, say a projective k-scheme.

The local moduli problem may then be phrased as follows. Describe the set of isomorphism classes of objects X' occuring as "arbitrary small deformations" of X.

In practice this means to define a natural filtration on the set of these isomorphism classes, such that each subset of the filtration may be given an algebraic structure, say as a k-scheme or, more generally, as an algebraic space. We shall refer to any such, natural, filtration $\{M_\tau\}$ as a <u>local moduli suite</u> of X.

This done, one would like to find the local structure of these new objects, their dimensions etc.

Our approach starts with a study of the formal moduli $\underline{H}^\wedge = \mathrm{Spf}(H^\wedge)$ of X, see §1 and §2.

Recall that the tangent space of $\underline{H}^\wedge$ is isomorphic to $A^1(k,X,\mathcal{O}_X)$, the first generalized André-Quillen cohomology of X, see e.g. [La 1]. (We are painfully aware of the fact that these cohomology groups usually are denoted by $T(X)$, and are refered to as the cohomology of the cotangent complex. There are, however, good reasons not to adhere to this practice. Just look into the deformation theory for modules or Lie algebras.)

Put $\tau(X) = \dim A^1(k,X;\mathcal{O}_X)$.

In §1 we prove that there is a unique maximal closed sub-proscheme $\underline{H}_0^\wedge = \mathrm{Spf}(H_0^\wedge)$ of $\underline{H}^\wedge$ for which the obvious composition of natural transformations

$$\mathrm{Mor}(H_0^\wedge,-) \to \mathrm{Mor}(H^\wedge,-) \to \mathrm{Def}_X$$

is injective. $\underline{H}_0^\wedge$ is the <u>prorepresenting substratum</u> of $\underline{H}^\wedge$.

In §2 we study the group functor $\underline{I}(\pi^\wedge)$ of those automorphisms of $H^\wedge$ that leave the formal versal family invariant, and we prove that $\underline{H}_0^\wedge$ is the fixed proscheme of the action of its Lie algebra $\ell(\pi^\wedge)$, on $\underline{H}^\wedge$.

Consider now the following conditions, see the introduction to §3,

(A_1) There exists an algebraization

$$\pi: \tilde{X} \to \underline{H} = \mathrm{Spec}(H)$$

of the formal versal family

(V) π is formally versal (see definition (3.6)).

(A_2) π is versal in the étale topology.

In §3 we prove that, under the condition (A_1) and a set of conditions, (V'), implying in particular (V) and the smoothness of $\underline{H}$, there exists a <u>prorepresenting substratum</u> $\underline{H}_0$ of $\underline{H}$, the formalization of which at each closed point $\underline{t} \in \underline{H}_0$ is the corresponding prorepresenting substratum of the formal moduli of $X(\underline{t}) = \pi^{-1}(\underline{t})$.

In fact, let V denote the kernel of the Kodaira-Spencer map

$$g: \operatorname{Der}_k(H) \to A^1(H,\tilde{X};O_{\tilde{X}})$$

then V is a k-sub Lie algebra of Der(H), such that the formalization coincides with $\ell(\pi^\wedge)$. $\underline{H}_0$ is defined by the vanishing of V. Using this we prove, by glueing together the prorepresenting substrata of the versal bases corresponding to the various fibers of π, the <u>Theorem (3.18)</u> which asserts the existence of a local moduli suite $\{\underline{M}_\tau\}$, in the category of algebraic spaces, provided (A_1), (V') and (A_2) hold.

Put for every $\underline{t} \in \underline{H}$, $\tau(\underline{t}) = \tau(X(\underline{t}))$. Then the main result of §3 may be phrased as follows, see <u>Theorem (3.24)</u>: Let $\underline{S}_\tau$ be the τ-constant substratum of $\underline{H}$. Then there exists a scheme theoretic quotient of an open dense subscheme $\underline{U}_\tau$ of the normalization of $\underline{S}_\tau$ by the action of the Lie algebra V, and a quasifinite dominant morphism $\underline{U}_\tau/V \to \underline{M}_\tau$.
The rest of this monograph is concerned with hypersurfaces and hypersurface singularities, see §§4-7.

To illustrate the main ideas of these paragraphs, let us consider a plane affine curve defined by $f \in k[x_1,x_2]$, with only isolated singularities.

Put $X = \operatorname{Spec}(k[x_1,x_2]/(f))$. The cohomology $A^i = A^i(k,X;O_X)$ is, in this case $A^0 = \operatorname{Der}_k(k[x_1,x_2]/(f))$, $A^1 = k[x_1,x_2]/(f,\frac{\partial f}{\partial x_1},\frac{\partial f}{\partial x_2})$, $A^i = 0$ for $i \geq 2$. The formal moduli is therefore $H^\wedge = k[[t_{\underline{\alpha}}]]_{\underline{\alpha}\in I}$ where $I \subseteq Z_+^2$ is such that $\{x_1^{\alpha_1}x_2^{\alpha_2}\}_{(\alpha_1,\alpha_2)\in I}$ is a basis for A^1.

There exists an algebraization

$$\pi: \tilde{X} \to \underline{H}$$

of the formal versal family of X, with $\underline{H} = \mathrm{Spec}(H)$, $H = k[t_{\underline{\alpha}}]_{\underline{\alpha}\in I}$,
$\tilde{X} = \mathrm{Spec}(H[x_1,x_2]/(F))$, $F = f + \sum_{\underline{\alpha}\in I} t_{\underline{\alpha}} x_1^{\alpha_1} x_2^{\alpha_2}$.

Thus (A_1) is satisfied. Since $\tilde{X}$ is also a hypersurface F, we find:

(2) $A^1(H,\tilde{X};O_{\tilde{X}}) = H[x_1,x_2]/(F,\frac{\partial F}{\partial x_1},\frac{\partial F}{\partial x_2})$ is an H-module of finite type

(3) $A^i(H,\tilde{X};O_{\tilde{X}}) = 0$ for $i \geq 2$

Together with the obvious,

(1) H is k-smooth

these properties show that the conditions (V') refered to above, are satisfied.

However, the condition (A_2) is not in general satisfied. In fact, as we know, the family $x^3+y^2+t_1x+t_0$, $27t_0^2+4t_1^3 \neq 0$, is not constant even though every fiber is smooth and therefore rigid.

Consequently we may not use the technique of §3 to produce a local moduli suite for X. The conclusions of (3.18) and (3.24) are, in fact, false in the affine case.

If, however, instead of the affine scheme $\mathrm{Spec}(k[x_1,x_2]/(f))$, we consider the formal scheme

$$X = \mathrm{Spf}(k[[x_1,x_2]]/(f))$$

we are in a much better situation.

The correct cohomology in this case is easily computed. We find

$$A^0 = \mathrm{Der}_k^c(k[[x_1,x_2]]/(f))$$

$$A^1 = (x_1,x_2)/(x_1,x_2)\cdot(\frac{\partial f}{\partial x_1},\frac{\partial f}{\partial x_2})+(f)$$

$$A^i = 0\ ,\quad i \geq 2$$

As in the affine case, there is an algebraization of the formal moduli, with H and F defined by the same formulas as above. The only difference is that now all the conditions (A_1), (V') and (A_2) are satisfied. Therefore we may apply the method of §3 to produce a local moduli suite for X.

Consider the Kodaira-Spencer morphism

$$g\colon \mathrm{Der}_k(H) \to A^1(H,\tilde{X};O_{\tilde{X}})$$

$g(\frac{\partial}{\partial t_{\underline{\alpha}}})$ is the class of $\frac{\partial F}{\partial t_{\underline{\alpha}}}$ in

$$A^1(H,\tilde{X};O_{\tilde{X}}) = (x_1,x_2)\cdot H[[x_1,x_2]]/(x_1,x_2)(\frac{\partial F}{\partial x_1},\frac{\partial F}{\partial x_2})+(F).$$

The kernel V of g is easily seen to be a k-Lie subalgebra of $\mathrm{Der}_k(H)$.

Let $\underline{H}_o := \underline{H}_o(f)$ be the closed subscheme of $\underline{H}$ on which V vanishes. $\underline{H}_o$ is the <u>prorepresenting substratum</u> of $\underline{H}$ and forms the "inner room", $\underline{M}_{\tau(f)}$ of the local moduli suite of X.

Recall the definitions of the Tjurina number

$\tau(f) = \dim_k(k[[x_1,x_2]]/(f,\frac{\partial f}{\partial x_1},\frac{\partial f}{\partial x_2}))$ and, the Milnor number $\mu(f) = \dim_k(k[[x_1,x_2]]/(\frac{\partial f}{\partial x_1},\frac{\partial f}{\partial x_2}))$, associated to the singularity f.

Consider for every τ, $0 \leq \tau \leq \tau(f)$, the subscheme of $\underline{H}$

$$\underline{S}_\tau = \{\underline{t}\in\underline{H} \mid \tau(F(\underline{t}))=\tau\}$$

<u>Theorem (3.18)</u> implies the existence of a cartesian square in the category of algebraic spaces

$$(0)\qquad \begin{array}{ccc} \tilde{X}_\tau & \longrightarrow & \mathcal{H}_\tau \\ \pi_\tau \downarrow & \square & \downarrow \bar{\pi}_\tau \\ \underline{S}_\tau & \xrightarrow{\sigma_\tau} & \underline{M}_\tau \end{array}$$

where π_τ is the restriction of π to $\underline{S}_\tau$, and $\underline{M}_\tau$ is a glueing together of the prorepresenting substrata $\underline{H}_0(F(\underline{t}))$ for $\underline{t} \in \underline{S}_\tau$. The family $\{\underline{M}_\tau\}_{0\leq\tau\leq\tau(f)}$ is, by definition, the <u>local moduli suite of X (or f)</u>.

This notion is justified by the following facts

(1) $\bar{\pi}_\tau$ is flat

(2) If the restriction of $\bar{\pi}_\tau$ to a connected subscheme $\underline{K} \to \underline{M}_\tau$ is constant, then $\underline{K}$ is a closed point.

<u>Theorem (3.24)</u> implies that there exists an open dense subscheme $\underline{U}_\tau$ of the normalization of $\underline{S}_\tau$, and a geometric quotient $\underline{N}_\tau = \underline{U}_\tau/V_\tau$ together with a dominant quasifinite morphism

$\underline{N}_\tau \to \underline{M}_\tau$,

Here we use the fact that $\underline{S}_\tau$ is stable under V, and we have put $V_\tau = V|S_\tau$.

The example $f = x_1^5+x_2^{11}$, treated extensively in §5 and §7, shows that we may not in general assume $\underline{U}_\tau = \underline{S}_\tau$. In fact, in this example $\underline{S}_{35}/V_{35}$ does not exist as a geometric quotient.

In §§4 to 7 we restrict to the quasihomogeneous case, and we change our setting slightly by fixing the topological type of the deformations.

Let $\underline{H}_\mu = \{\underline{t}\in\underline{H}|\mu(F(\underline{t}))=\mu(f)\}$, and put $\underline{S}_{\mu\tau} = \underline{S}_\tau \cap \underline{H}_\mu$. Notice that $\underline{S}_{\mu,\tau_{min}}$ is open and dense in $\underline{H}_\mu$, where $\tau_{min} = \min\{\tau(F(\underline{t}))|\underline{t}\in\underline{H}_\mu\}$. Put also $\underline{M}_{\mu,\tau} = \sigma_\tau(\underline{S}_{\mu\tau})$. Then $\{\underline{M}_{\mu,\tau}\}_{\tau_{min}\leq\tau\leq\tau(f)}$ is the <u>local μ-constant moduli suite of X</u>.

In §4 we compute the restriction of V to $\underline{H}_\mu$. Using this we are able in §5 to compute the dim $\underline{M}_{\mu,\tau_{min}}$, and to prove

<u>Theorem (5.1)</u>: If $f = x_1^{a_1}+ x_2^{a_2}$ then:

$$\dim \underline{M}_{\mu,\tau_{min}} \geq\ldots\geq \dim \underline{M}_{\mu,\tau} \geq \dim \underline{M}_{\mu,\tau+1} \geq\ldots\geq \dim \underline{M}_{\mu,\tau(f)}$$

This result shows in particular, that the generic μ-constant deformation of f represents (a component of) the generic isomorphism class of deformations of f. The example $f = x_1^3+x_2^{10}+x_3^{19}$, treated in §5, shows that the above result is not true in higher dimensions!

In §6 we turn to the study of $\underline{M}_{\mu,\tau_{min}}$. The main result is:

<u>Theorem (6.1)</u>. Let $f = x_1^{a_1}+x_2^{a_2}$ with $(a_1,a_2) = 1$, then there exists a geometric quotient $\underline{T}_{\tau_{min}} = \underline{S}_{\mu,\tau_{min}}/V$ which is a coarse moduli scheme for plane curve singularities with semigroup $\Gamma = \langle a_1,a_2\rangle$ and minimal Tjurina number.

This result has been generalized to any quasihomogeneous plane curve singularity, by B. Martin and these authors, see [L-M-P].

Finally, in §7 we treat the example $f = x^5 + y^{11}$ in detail, and in §8 we add an algorithm for computing the Kodaira-Spencer kernel V for plane curves.

The main results are also summed up in the introductions to each paragraph.

This monograph is the outgrowth of a collaboration between the two authors during the last 5 years. A first, very sketchy version appeared in 1983 [La-Pf].

Many authors have previously treated the same subject, see e.g. Merle [Me], Teissier [Z], Washburn [Wash], and Zariski [Z]. In particular Zariski in his lecture notes [Z], published in 1973, laid the foundations to the study of hypersurface singularities in the algebroid sense. Obviously, his results have influenced upon our work, even though our methods are quite different, and our goals seemingly somewhat wider.

It should be explicitly mentioned that Palamodov in [Pal] studied the notion of prorepresenting substratum (§1), and that Saito in [S], has the same calculation as we obtain in §4, of the kernel of the Kodaira-Spencer map. These ideas are, however, part of the folklore of the last decades, nourished by the work of Grothendieck, Mumford, Artin and Schlessinger, and were at the origin of one of the authors interest in this subject, see [La1].

We have profited a lot on discussions with our colleages at Berlin and Oslo.

In particular we would like to thank Dr. Bernd Martin for his contribution to §§6-8.

We are deeply indebted to Randi Møller and Inger Jansen for their skill and really infinite patience in typing the manuscript.

Finally, we are grateful for the financial support by the Humboldt-Universität zu Berlin, DDR, by the University of Oslo, Norway, and by the Norwegian Research Council, NAVF.

Given an algebraically closed field k of characteristic zero, we shall, throughout §1-§3 be concerned with an algebraic object X such as

<u>Example 1</u>. $X = \underline{c}$, a small category of k-schemes. Put

$$A^i = A^i(k, \underline{c}, O_{\underline{c}}), \quad i \geq 0, \text{ see [La 1].}$$

Example 2. $X = \mathrm{Spec}(A)$, A any k-algebra with isolated singularities. In particular, we shall be interested in the case where $A = k[X_1,\dots,X_n]/(f)$ is a hypersurface. In this case $A^0 = \mathrm{Der}_k(k[X_1,\dots,X_n]/(f))$, $A^1 = k[X_1,\dots,X_n]/(\frac{\partial f}{\partial X_1},\dots,\frac{\partial f}{\partial X_n}, f)$ and $A^2 = 0$. For the notion of hypersurface singularity, see §4.

Example 3. $X = \underline{e}$, a small category of O_Y-Modules where Y is some k-scheme. Here $A^i = \mathrm{Ext}^i_{O_Y}(O_{\underline{e}}, O_{\underline{e}})$, $i \geq 0$ are defined as in [La 1] with Hom replacing Der. See the concluding remark, loc.cit. p. 150, of [La 1] and [La 2].

Example 4. $X = E$, a coherent $O_{\underline{P}^n}$-Module. $A^i = \mathrm{Ext}^i_{O_{\underline{P}^n}}(E,E)$, $i \geq 0$. Of particular interest is the case where E is a vector bundle on P^n.

Assume now that $\dim_k A^i < \infty$ for $i = 1,2$. Then, see e.g. [La 1], (4.2.4), there exist in all these cases a formal moduli $\hat{H}$ (a prorepresenting hull for the deformation functor) of X, and a formal versal family

$$\hat{\pi}: \hat{X} \to \mathrm{Spf}(\hat{H}) = \underline{\hat{H}} .$$

The first part of this monograph , §§1-2 is devoted to the study of $\hat{\pi}$ in this generality.

§1. THE PROREPRESENTING SUBSTRATUM OF THE FORMAL MODULI

Introduction. Let X be an object of the type considered in the main introduction above.

The basic notion in the study of local moduli of X is the notion of prorepresenting substratum of the formal moduli.

If $H^\wedge$ is the formal moduli of X, then the prorepresenting substratum $H_0^\wedge$ is the unique maximal quotient of $H^\wedge$ for which the obvious composition

$$\mathrm{Mor}(H_0^\wedge,-) \to \mathrm{Mor}(H^\wedge,-) \to \mathrm{Def}_X$$

is injective.

This quotient exists in all generality, and the object of this § is its construction, see (1.3).

In §3, we shall want to extend this notion to the algebraization of the formal moduli. It turns out that this is facilitated by the introduction of the concept of the n-th equicohomological substratum $H_{(n)}^\wedge$ of $H^\wedge$, see (1.5), and by proving, (1.6), that $H_0^\wedge$ coincides with the 0-th equicohomological substratum $H_{(0)}^\wedge$.

Let X be any algebraic object of the type discussed in the Introduction, and consider the deformation functor

$$\mathrm{Def}_X\colon \underline{\ell} \to \underline{\mathrm{Sets}},$$

the corresponding cohomology $A^i = A^i(k,X;O_X)$, $i \geq 0$ and the universal obstruction morphism

$$o_X\colon T^2 \to T^1$$

where $T^i = \mathrm{Sym}_k(A^{i*})^\wedge$. Denote by

$$H^\wedge = T^1 \underset{T^2}{\otimes} k$$

the formal moduli of X, i.e. the (prorepresenting) hull of the deformation functor Def_X, and put

$$\underline{H}^{\wedge} = \mathrm{Spf}(H^{\wedge}).$$

In general there are lots of infinitesimal automorphisms of X, and non trivial obstructions for lifting these (see [Sch]). Therefore $H^{\wedge}$ does not necessarily prorepresent Def_X. However, as we shall see, there is a universal prorepresenting substratum $\underline{H}_0^{\wedge}$ of $\underline{H}^{\wedge}$, corresponding to a quotient

$$H_0^{\wedge} = H^{\wedge}/\mathfrak{a}$$

of $H^{\wedge}$.

In fact, let us consider the category $\underline{\ell}_H$ of all artinian local $H^{\wedge}$-algebras with residue field k.

Let $X^{\wedge}$ be the formal versal family on $H^{\wedge}$ defined by the identity element $1_{H^{\wedge}} \in \mathrm{Mor}(H^{\wedge}, H^{\wedge})$ and consider the functor

$$\underline{\mathrm{aut}}_{X^{\wedge}}: \underline{\ell}_H \to \underline{\mathrm{gr}}$$

defined by:

$$\underline{\mathrm{aut}}_{X^{\wedge}}(S) = \{\phi \in \mathrm{Aut}_S(X^{\wedge} \otimes_{H^{\wedge}} S) \mid \phi \otimes_S k = 1_X\} =: \mathrm{aut}_S(X^{\wedge} \otimes_{H^{\wedge}} S)$$

Theorem (1.1). Assume $\dim_k A^i$ is countable $i = 0,1$. Then there exists a morphism of complete local $H^{\wedge}$-algebras

$$o_{\underline{a}}: H \hat{\otimes} T^1 \to H \hat{\otimes} T^0$$

such that

(i) $\quad o_{\underline{a}}(\underline{m}_{H\hat{\otimes}T^1}) \subseteq \underline{m}^2_{H\hat{\otimes}T^0}$

(ii) $\quad a_{X^{\wedge}} := (H\hat{\otimes}T^0) \otimes_{H\hat{\otimes}T^1} H^{\wedge}$

is a prorepresenting hull for the functor $\underline{\mathrm{aut}}_{X^{\wedge}}$.

Proof. This follows from the proof of [La 1], (4.2.4) with $\underline{\mathrm{aut}}_{X^{\wedge}}$ replacing Def_X and A^{i-1} replacing A^i, $i = 1,2$. Q.E.D.

Recall that there is the usual automorphism functor of X,

$$\underline{\mathrm{Aut}}_X : \underline{\mathrm{sch}}/k \to \underline{\mathrm{gr}}$$

defined by:

$$\underline{\mathrm{Aut}}_X(\underline{S}) = \mathrm{Aut}_{\underline{S}}(X \times \underline{S})$$

Assume $\underline{\mathrm{Aut}}_X$ is represented by the k-scheme $\mathrm{Aut}(X)$ and let $1 \in \mathrm{Aut}(X)$ be the identity element. Then the completion $\hat{O}_{\mathrm{Aut}(X),1}$ of the local ring of $\mathrm{Aut}(X)$ at 1, represents the fiber-functor of $\underline{\mathrm{Aut}}_X$ at $1 \in \mathrm{Aut}_k(X)$, i.e. the functor

$$\underline{\mathrm{aut}}_X : \underline{\ell} \to \underline{\mathrm{gr}}$$

defined by

$$\underline{\mathrm{aut}}_X(S) = \{\phi \in \mathrm{Aut}_S(X \otimes S) \mid \phi \otimes_S k = 1\} =: \mathrm{aut}_S(X \otimes S)$$

Let a_X be the prorepresenting hull of $\underline{\mathrm{aut}}_X$, such that with the assumption above

$$a_X \simeq \hat{O}_{\mathrm{Aut}(X),1}.$$

Notice that if $\mathrm{Aut}(X)$ is smooth, then $a_X \simeq \mathrm{Sym}_k(A^{0*})^\wedge$ (see [La1] Ch. 4).

<u>Definition (1.2)</u>. Let the ideal $\mathfrak{a} \subseteq H^\wedge$ be generated by the coefficiants of the elements of $o_{\underline{a}}(\underline{m}) \subset H \hat{\otimes} T^0$, $\underline{m}$ being the maximal ideal of $H \hat{\otimes} T^1$. Then the <u>prorepresenting substratum</u>

$$\underline{H}_0^\wedge \subseteq \underline{H}^\wedge$$

is the formal subscheme defined by $\mathfrak{a}$.

Put $H_0^\wedge = H^\wedge/\mathfrak{a}$. Then $\underline{H}_0^\wedge = \mathrm{Spf}(H_0^\wedge)$ and we shall, mildly abusing the notations, also speak about the prorepresenting substratum $H_0^\wedge$.

By construction of $o_{\underline{a}}$ it is clear that $H_0^\wedge$ is the maximal quotient of $H^\wedge$ for which

$$a_{X^\wedge} \otimes_{H^\wedge} H_0^\wedge$$

is $H_0^\wedge$-smooth.

<u>Proposition (1.3)</u>. $H_0^\wedge$ is the maximal quotient of $H^\wedge$ for which the canonical morphism of functors on $\underline{\ell}$,

$$\rho_0 : \mathrm{Mor}(H_0^\wedge, -) \to \mathrm{Def}_X$$

is injective.

<u>Proof</u>. Let $H_1^\wedge$ be a quotient of $H^\wedge$, and assume $\psi_1, \psi_2 \in \mathrm{Mor}(H_1^\wedge, R)$ are mapped onto the same element $\bar{\psi}_1 = \bar{\psi}_2$ in $\mathrm{Def}_X(R)$. This, of course, means that there exists an R-isomorphism $X^\wedge \underset{\psi_1}{\otimes} R \overset{\Phi}{\simeq} X^\wedge \underset{\psi_2}{\otimes} R$ where at the left side R is considered as $H^\wedge$-module via ψ_1 and at the right hand side R is considered as $H^\wedge$-module via ψ_2.

We may assume, by induction, $\psi_1 \equiv \psi_2 \pmod{\underline{n}}$ where $\underline{n}$ is some ideal of R killed by the maximal ideal $\underline{m}_R$. Then $\Phi \otimes R/\underline{n}$ is an automorphism of $\hat{X} \otimes_{H^\wedge} R/\underline{n}$, corresponding to a morphism $a_{X^\wedge} \otimes_{H^\wedge} H_1^\wedge \to R/\underline{n}$. If $a_{X^\wedge} \otimes_{H^\wedge} H_1^\wedge$ is formally $H_1^\wedge$-smooth, then obviously this morphism may be lifted to a morphism $a_{X^\wedge} \otimes_{H^\wedge} H_1^\wedge \to R$, proving that $\Phi \otimes_R R/\underline{n}$ is liftable as an automorphism to some $\Phi_1 : X^\wedge \underset{\psi_2}{\otimes} R \tilde{\to} X^\wedge \underset{\psi_2}{\otimes} R$. But then $\Phi \circ \Phi_1^{-1} : X^\wedge \underset{\psi_1}{\otimes} R \tilde{\to} X^\wedge \underset{\psi_2}{\otimes} R$ is an isomorphism extending the identity of $X^\wedge \otimes_{H^\wedge} R/\underline{n}$. Thus $\psi_1 = \psi_2$. From this follows that $\rho_0 : \mathrm{Mor}(H_0^\wedge, -) \to \mathrm{Def}_X$ is injective.

Conversely assume $H_1^\wedge$ is a quotient of $H^\wedge$ such that $\rho_1 : \mathrm{Mor}(H_1^\wedge, -) \to \mathrm{Def}_X$ is injective. If R, an object of $\underline{\ell}_H$, is an $H_1^\wedge$-algebra, then any automorphism $\Phi_{\underline{n}}$ of $X^\wedge \otimes_{H^\wedge} R/\underline{n} = (X^\wedge \otimes_{H^\wedge} H_1^\wedge) \otimes_{H_1^\wedge} R/\underline{n}$ may always be lifted to an automorphism of $X^\wedge \otimes_{H^\wedge} R$. It follows that $a_{X^\wedge} \otimes_{H^\wedge} H_1^\wedge$ has to be formally smooth, which proves the proposition.

Q.E.D.

<u>Remark (1.4)</u>. Recall that $H^\wedge/\underline{m}^2$ represents the restriction of the deformation functor Def_X to the subcategory $\underline{\ell}_2 = \{R \in \underline{\ell} \mid \underline{m}_R^2 = 0\}$ of

$\underline{\ell}$. Notice that, nevertheless, $H/\underline{m}^2$ is rarely a quotient of $\hat{H}_0$, see §4 for lots of examples.

Consider for any $n \geq 0$ the subfunctors Def_X^n of Def_X defined by:

$$\mathrm{Def}_X^n(R) = \{X_R \in \mathrm{Def}_X(R) \,|\, A^n(R,X_R;O_{X_R}) \text{ is a deformation of } A^n(k,X;O_X)\}$$

Then one may prove that Def_X^n has a (prorepresenting) hull $\hat{H}_{(n)}$ which is a quotient of $\hat{H}$.

Definition (1.5). The formal subscheme $\mathrm{Spf}\,\hat{H}_{(n)}$ is called the n-th equicohomological substratum of $\underline{\hat{H}}$, and is denoted by $\underline{\hat{H}}_{(n)}$.

Proposition (1.6). The prorepresenting substratum $\hat{H}_0$ coincides with the 0-th equicohomological substratum $\hat{H}_{(0)}$.

Proof. For any object R of $\underline{\ell}_H$, there exists a bijective map

$$\exp : \ker\{A^0(R,\hat{X}\otimes_{\hat{H}}R;O_{\hat{X}\otimes_{\hat{H}}R}) \to A^0(k,X;O_X)\} \to \mathrm{aut}_R(\hat{X}\otimes_{\hat{H}}R)$$

the inverse of which is log.

In fact, any element σ of $\ker\{A^0(R,\hat{X}\otimes_{\hat{H}}R;O_{\hat{X}\otimes_{\hat{H}}R}) \to A^0(k,X;O_X)\}$ is an element of $\mathrm{Der}_R(O_{\hat{X}\otimes_{\hat{H}}R})$ (resp. of $\mathrm{End}_R(O_{\hat{X}\otimes_{\hat{H}}R})$ in the module case) mapping any local section x of $O_{\hat{X}\otimes_{\hat{H}}R}$ into $\underline{m}_R\cdot(O_{\hat{H}\otimes_{\hat{H}}R})$.

Since for some n, $\underline{m}_R^n\cdot(O_{\hat{X}\otimes_{\hat{H}}R}) = 0$, $\exp\sigma$ is defined. Given a homomorphism $\nu : \hat{H}_{(0)} \to R$, then the map

$$\eta_S^R : A^0(R,\hat{X}\otimes_{\hat{H}}R;O_{\hat{X}\otimes_{\hat{H}}R}) \to A^0(S,\hat{X}\otimes_{\hat{H}}S;O_{\hat{X}\otimes_{\hat{H}}S})$$

induced by a surjective morphism $\pi : R \to S$ in $\underline{\ell}$, is also surjective. This implies that the corresponding

$$\delta_S^R : \mathrm{aut}_R(\hat{X}\otimes_{\hat{H}}R) \to \mathrm{aut}_S(\hat{X}\otimes_{\hat{H}}S)$$

is surjective, thus by definition of $\hat{H}_0$ there exists a unique morphism $\bar{\mu} : \hat{H}_0 \to \hat{H}_{(0)}$, such that the diagram

$$\begin{array}{ccc} H^\wedge & \rightarrow & H^\wedge_{(0)} \\ \downarrow & \overset{\bar{\mu}}{\nearrow} & \downarrow \nu \\ H^\wedge_0 & \xrightarrow{\mu} & R \end{array}$$

commutes.

On the other hand, given a homomorphism $\mu: H^\wedge_0 \rightarrow R$ we may compose it with the canonical homomorphism $R \rightarrow R[\varepsilon]$ to obtain a homomorphism $H^\wedge_0 \rightarrow R[\varepsilon]$. By definition of $H^\wedge_0$ it follows that for any surjective $\pi: R \twoheadrightarrow S$ in $\underline{\ell}$ the horizontal maps in the following diagram are surjective,

$$\begin{array}{ccc} \operatorname{aut}_{R'}(X^\wedge \otimes_{H^\wedge} R') & \xrightarrow{\delta^{R'}_{S'}} & \operatorname{aut}_{S'}(X^\wedge \otimes_{H^\wedge} S') \\ \downarrow \delta^{R'}_{R} & & \downarrow \delta^{S'}_{S} \\ \operatorname{aut}_{R}(X^\wedge \otimes_{H^\wedge} R) & \xrightarrow{\delta^{R}_{S}} & \operatorname{aut}_{S}(X^\wedge \otimes_{H^\wedge} S) \end{array}$$

where we have put $R' = R[\varepsilon]$, $S'=S[\varepsilon]$. Since the vertical maps have sections, we find that

$$A^0(R, X^\wedge \otimes_{H^\wedge} R; O_{H^\wedge} \otimes_{H^\wedge} R) = \ker \delta^{R'}_{R}$$

maps surjectively onto

$$A^0(S, X^\wedge \otimes_{H^\wedge} S; O_{X^\wedge} \otimes_{H^\wedge} S) = \ker \delta^{S'}_{S}.$$

By definition of $H^\wedge_{(0)}$ there exists a unique morphism $\underline{\mu}: H^\wedge_{(0)} \rightarrow H^\wedge_0$ such that the diagram

$$\begin{array}{ccc} H^\wedge & \rightarrow & H^\wedge_{(0)} \\ \downarrow & \overset{\underline{\mu}}{\swarrow} & \downarrow \nu \\ H^\wedge_0 & \xrightarrow{\mu} & R \end{array}$$

commutes. Consequently we find $H^\wedge_0 = H^\wedge_{(0)}$.

Q.E.D.

Remark (1.7). Let $\underline{H}^{\wedge}_{(i)}$ be the i-th equicohomological substratum of $\underline{H}^{\wedge}$. Let $\underline{H}_{\infty}$ be the intersection of the $\underline{H}^{\wedge}_{(i)}$'s, and let $X^{\wedge}_{\infty} \to \underline{H}_{\infty}$ be the restriction of π to $\underline{H}_{\infty}$. Then $A^i(H^{\wedge}_{\infty}, X^{\wedge}_{\infty}; O_{X^{\wedge}_{\infty}})$ is H_{∞}- flat for all $i>0$.

Suppose that $A^i(H_{\infty}, X^{\wedge}_{\infty}; O_{X^{\wedge}_{\infty}})$ is of finite type over H_{∞}. Then, in particular, $A^i(H_{\infty}/\underline{m}^n_x; X^{\wedge}_{\infty} \otimes_{H_{\infty}} H_{\infty}/\underline{m}^n)$ is reflexive as an $H_{\infty}/\underline{m}^n_x$-module, for all $n>0$.

Now assuming we have a flat family $\eta: Y \to \mathrm{Spec}(S)$ such that $A^i = A^i(S, Y; O_Y)$ is reflexive as an S-module for $i = 1,2$, there exist a morphism of complete S-algebras

$$T^2_S = \mathrm{Sym}_S(A^{2*})^{\wedge} \to T^1_S = \mathrm{Sym}_S(A^{1*})^{\wedge}$$

such that the k-algebra

$$(T^1_S \mathbin{\hat{\otimes}}_{T^2_S} S) \otimes_S k(\underline{s})$$

is the formal moduli of $Y(\underline{s}) = \eta^{-1}(\underline{s})$ for all closed points $\underline{s} \in \mathrm{Spec}(S)$. The proof of this parallels the proof of [La1]; (4.4.2).

§2. AUTOMORPHISMS OF THE FORMAL MODULI

Introduction. The prorepresenting substratum constructed in §1 is a closed subproscheme $\underline{H}_0^{\wedge}$ of $\underline{H}^{\wedge}$. In this § we shall show that $\underline{H}_0^{\wedge}$ is the fixed proscheme of $\underline{H}^{\wedge}$ under the action of a subgroup functor $\underline{i}_X$ of $\underline{\text{aut}}_{H^{\wedge}}$ contained in the covering automorphism group functor of the morphism $\rho: \text{Mor}(H^{\wedge},-) \to \text{Def}_X$.

Notice that this does not imply that $\underline{H}_0^{\wedge}$ is the fixed proscheme of some natural subgroup of $\text{aut}_k(\underline{H}^{\wedge})$, see (1.4).

Since the group-functor $\underline{i}_X$ does not, in general, extend to a group-functor of automorphisms of an algebraization H of $H^{\wedge}$, we are led to consider the Lie algebra $\ell(\pi^{\wedge}) = \text{lie}\ \underline{i}_X$.

The main result of this § is (2.5), where we, in particular, prove that $\ell(\pi^{\wedge})$ is an $H^{\wedge}$-submodule and a sub k-Lie algebra of $\text{Der}_k(H^{\wedge})$, such that $\ell(\pi^{\wedge}) \otimes_{H^{\wedge}} k \simeq A^0(k,X;O_X)/A^0_{\pi^{\wedge}}$ where $A^0_{\pi^{\wedge}}$ is the Lie ideal in $A^0(k,X;O_X)$ of those infinitesimal automorphisms of X that can be lifted to $X^{\wedge}$. Moreover $\ell(\pi^{\wedge}) \otimes_{H^{\wedge}} H_0^{\wedge}$ is an $H_0^{\wedge}$-Lie algebra defining a deformation of the Lie algebra $L(X) = A^0(k,X;O_X)/A^0_{\pi^{\wedge}}$ to $H_0^{\wedge}$.

Consider any formal deformation of X,

$$\pi: Y^{\wedge} \to \underline{S}^{\wedge} = \text{Spf}(S^{\wedge}).$$

In particular we shall be interested in the formal versal family

$$\pi^{\wedge}: X^{\wedge} \to \underline{H}^{\wedge} = \text{Spf}(H^{\wedge}).$$

Let $\underline{\text{aut}}_{S^{\wedge}}$ be the subfunctor of $\underline{\text{Aut}}_{S^{\wedge}}: \underline{\ell} \to \underline{\text{Sets}}$ such that for every object R of $\underline{\ell}$, $\underline{\text{aut}}_{S^{\wedge}}(R)$ is the subset of those $\psi \in \underline{\text{Aut}}_{S^{\wedge}}(R)$ for which the following diagram commutes

$$\begin{array}{ccc} S^{\wedge} \otimes R & \xrightarrow{\phi} & S^{\wedge} \otimes R \\ \downarrow & & \downarrow \\ R & \xrightarrow{id} & R\,. \end{array}$$

Consider the subfunctors $\underline{I}_\pi$, $\underline{i}_\pi$ and $\underline{i}^*_\pi$ of $\underline{\text{aut}}_{S^\wedge}$ defined by

$$\underline{I}_\pi(R) = \{\psi \in \underline{\text{aut}}_{S^\wedge}(R) \mid Y^\wedge \otimes R \overset{\chi}{\simeq} Y^\wedge \underset{\psi}{\otimes} (S^\wedge \otimes R)\}$$

$$\underline{i}_\pi(R) = \{\psi \in \underline{I}_k(R) \mid \chi \underset{S^\wedge \otimes R}{\otimes} k = id_X\}$$

$$\underline{i}^*_\pi(R) = \{\psi \in \underline{i}_\pi(R) \mid \chi \underset{S^\wedge}{\otimes} k = id_{X\otimes R}\}$$

where we have written $Y^\wedge \underset{\psi}{\otimes} (S^\wedge \otimes R)$ for the pull-back of $Y^\wedge \times \text{Spec}(R)$ by the morphism

$$\text{Spf}(\psi): \text{Spf}(S^\wedge \otimes R) \to \text{Spf}(S^\wedge \otimes R)\}.$$

In particular, corresponding to the formal versal family, we put

$$\underline{I}_X = \underline{I}_{\pi^\wedge}, \qquad \underline{i}_X = \underline{i}_{\pi^\wedge}$$

$$I(X) = \underline{I}_X(k) \qquad i(X) = \underline{i}_X(k).$$

Notice that $i(X) = \underline{i}_X(k) = \underline{i}^*_X(k)$, by definition, consists of those automorphisms ψ of $H^\wedge$ which leaves

$$\rho: \text{Mor}\,(\hat{H},-) \twoheadrightarrow \text{Def}_X(-)$$

fixed, i.e. s.t. the diagram

$$\begin{array}{ccc} \text{Mor}\,(H^\wedge,-) & \xrightarrow{\psi^*} & \text{Mor}\,(H^\wedge,-) \\ & {\scriptstyle\rho}\searrow \quad \swarrow{\scriptstyle\rho} & \\ & \text{Def}_X(-) & \end{array}$$

commutes. The group-functor $\underline{i}_X$ thus measures the extent of non pro-representability of Def_X.

Recall, from §1, that if $\underline{\text{Aut}}_X$ is smooth, then $\underline{\text{aut}}_X$ is prorepresented by T^0.

Now it is easy to show that $\underline{\text{aut}}_X$, restricted to $\underline{\ell}$, is smooth, in all generality. Let us prove it when X is a k-algebra.

Consider surjective morphisms $\eta: T \to R$ and $\mu: R \to S$ of $\underline{\ell}$ such that $\underline{m}_R \cdot \ker \mu = 0$.

Suppose $\theta_R \in \underline{\mathrm{aut}}_X(R)$ is such that $\theta_R \otimes S = \mathrm{id} \in \underline{\mathrm{aut}}_X(S)$. Then $\theta_R = \mathrm{id}_{X\otimes R} + D$ where $D \in \mathrm{Der}_k(X, X\otimes \ker\mu)$. We may write $D = \sum_{i=1}^{m} r_i D_i$, $r_i \in \ker \mu$, $D_i \in \mathrm{Der}_k(X)$. Pick $t_i \in T$ such that $\eta(t_i) = r$ and consider the derivation $D' = \sum_{i=1}^{m} t_i D_i \in \mathrm{Der}_k(X, X\otimes T)$. D' defines in an obvious way a derivation $D' \in \mathrm{Der}_T(X\otimes T)$. Let $\theta' = \exp D' = \mathrm{id} + D' + \frac{1}{2!}D'^2 + \frac{1}{3!}D'^3 + \ldots$. Then θ' is an element of $\mathrm{Aut}_T(X\otimes T)$ such that $\theta' \otimes_T R = \theta_R$. An easy induction argument then shows that $\underline{\mathrm{aut}}_X(T) \to \underline{\mathrm{aut}}_X(R)$ is surjective, thus $\underline{\mathrm{aut}}_X$ is smooth.

<u>Theorem (2.1)</u>. There is a (non-canonical) morphism of the underlying set-theoretical functors

$$\phi: \underline{\mathrm{aut}}_{X^\wedge}(H^\wedge \otimes -) \to \underline{\mathrm{Aut}}_{H^\wedge}$$

such that

(i) $\langle \mathrm{im}\phi \rangle = \underline{i}_X$, as subfunctors of $\underline{\mathrm{Aut}}_{H^\wedge}$.

(ii) $H_0^\wedge$ is the maximal quotient of $H^\wedge$ trivializing $\underline{i}_X$.

<u>Proof</u>. Consider the (prorepresenting) hull $a_X{}^\wedge$, of the group-functor $\underline{\mathrm{aut}}_X{}^\wedge$, see §1. The identity $\mathrm{id}: a_X{}^\wedge \to a_X{}^\wedge$ corresponds to the universal automorphism

$$\theta \in \mathrm{aut}_{a_X{}^\wedge}(X^\wedge \hat{\otimes}_{H^\wedge} a_X{}^\wedge) = \underline{\mathrm{aut}}_{X^\wedge}(a_X{}^\wedge)$$

By (1.1) $a_X{}^\wedge = H \hat{\otimes} T^0/\mathfrak{a}$, where $\mathfrak{a}$ is an ideal contained in the square of the maximal ideal of $H \hat{\otimes} T^0$.

Consider now the trivial lifting $X^\wedge \hat{\otimes} T^0$ of $X^\wedge \hat{\otimes}_{H^\wedge} a_X{}^\wedge$ to $H \hat{\otimes} T^0$ defined in terms of the quotient morphism $q: H \hat{\otimes} T^0 \to a_X{}^\wedge$, i.e. the lifting corresponding to the canonical homomorphism

$$i: H^\wedge \to H \hat{\otimes} T^0$$

The automorphism θ of $X^\wedge \hat{\otimes}_{H^\wedge} a_X{}^\wedge$ may be lifted to an isomorphism $\bar{\theta}$ making the following diagram commutative

$$\begin{array}{ccc} X^\wedge \hat{\otimes} T^0 & \xrightarrow[\bar{\theta}]{} & X^\wedge \otimes (H \hat{\otimes} T^0) \\ \downarrow & & \phi \downarrow \\ X^\wedge \hat{\otimes}_{H^\wedge} {}^a X^\wedge & \xrightarrow[\theta]{} & X^\wedge \hat{\otimes}_{H^\wedge} {}^a X^\wedge \end{array}$$

where $\phi\colon H^\wedge \to H \hat{\otimes} T^0$ is some homomorphism such that $q \circ \phi = q \circ i$. By definition $H_0^\wedge$ is thus the maximal quotient of H for which $i \otimes_{H^\wedge} 1_{H_0^\wedge} = \phi \otimes_{H^\wedge} 1_{H_0^\wedge}$.

Consider the map which associates to every $\alpha \in \mathrm{Mor}_k(T^0, H^\wedge \otimes R) = \mathrm{Mor}_R(T^0 \hat{\otimes} R, H^\wedge \otimes R)$ the composition

$$\phi(\alpha)\colon H^\wedge \otimes R \xrightarrow[\phi \otimes 1_R]{} (H \hat{\otimes} T^0) \otimes R \xrightarrow[1_{H^\wedge} \otimes \alpha]{} H^\wedge \otimes R$$

Since $1_{H^\wedge} \otimes \alpha$ maps $\mathfrak{a}$ into the square of the maximal ideal of $H^\wedge \otimes R$, it follows from $q \circ \phi = q \circ i$ that $\phi(\alpha)$ reduces to the identity on the tangent level. Therefore $\phi(\alpha) \in \mathrm{Aut}_R(H^\wedge \otimes R)$, and we obtain a map

$$\phi_R\colon \underline{\mathrm{aut}}_X(H^\wedge \otimes R) \to \underline{\mathrm{Aut}}_{H^\wedge}(R)$$

which, as one easily checks, is functorial. Furthermore, by construction,

$$\bar{\theta} \underset{(1_{H^\wedge} \otimes \alpha)}{\otimes} (H^\wedge \otimes R)\colon X^\wedge \otimes R \xrightarrow{\sim} X^\wedge \underset{\phi(\alpha)}{\hat{\otimes}} (H^\wedge \otimes R)$$

is an $H^\wedge \otimes R$-isomorphism, so we know $\phi(\alpha) \in \underline{i}_X(R)$. We need only prove that ϕ_R maps $\underline{\mathrm{aut}}_X(H^\wedge \otimes R)$ onto $\underline{i}_X(R)$. The rest is clear. Therefore the proof is reduced to proving the next proposition.

Q.E.D.

<u>Proposition (2.2)</u>. (i) Let $\delta \in \mathrm{Mor}(H^\wedge, S)$ correspond to the deformation $X_S \in \mathrm{Def}_X(S)$. Then for every $\psi \in i(X)$, the morphism $\psi \circ \delta \in \mathrm{Mor}(H, S)$ corresponds to the same deformation X_S.

(ii) If the surjections $\rho_i\colon H^\wedge \otimes R \to S$, $i=1,2$, correspond to the same deformation X_S, then there exists a sequence of automorphisms $\alpha_n \in \underline{\mathrm{aut}}_X(H^\wedge \otimes R)$ such that $\rho_2 = \lim_{n \to \infty} \rho_1 \circ \phi(\alpha_2) \circ \ldots \circ \phi(\alpha_n)$.

Proof. (i) is obvious. To prove (ii) consider the morphisms $\rho_i^2 : H^\wedge \otimes R \to S/\underline{m}^2 = S_2$. By assumption, we have a commutative diagram

$$\begin{array}{ccc}
(X^\wedge \otimes R) \otimes_{\rho_1} S & \underset{\tau}{\tilde{\to}} & (X^\wedge \otimes R) \otimes_{\rho_2} S \\
\downarrow & & \downarrow \\
(X^\wedge \otimes R) \otimes_{\rho_1^2} S_2 & \underset{\tau_2}{\tilde{\to}} & (X^\wedge \otimes R) \otimes_{\rho_2^2} S_2 \\
\downarrow & & \downarrow \\
X & = & X
\end{array}$$

where τ and τ_2 are isomorphisms.

Since H_2 represents the deformation functor Def_X restricted to the subcategory $\underline{\ell}_2$ of $\underline{\ell}$, $\rho_1^2 = \rho_2^2$ and τ_2 is therefore an automorphism. As such it corresponds to a morphism $\mu_2 : a_X \to S_2$, which, composed with the canonical morphism $H \hat{\otimes} T^0 \to a_X$, gives us an $H^\wedge$-morphism $\bar{\mu}_2 : H \hat{\otimes} T^0 \to S_2$. Lift $\bar{\mu}_2$ to an $H^\wedge$-morphism $\alpha_2' : H \hat{\otimes} T^0 \to H^\wedge \otimes R$, and consider the composition

$$\phi(\alpha_2) : H^\wedge \otimes R \xrightarrow[\phi \otimes 1_R]{} (H \hat{\otimes} T^0) \otimes R \xrightarrow[1_H \otimes \alpha_2]{} H^\wedge \otimes R$$

where α_2 is the composition of $T^0 \to H \hat{\otimes} T^0$ and α_2'.

By construction there is a commutative diagram

$$\begin{array}{ccc}
X^\wedge \otimes R & \overset{\bar{\theta} \otimes_{\phi(\alpha_2)} (H^\wedge \otimes R)}{\underset{\sim}{\to}} & (X^\wedge \otimes R) \otimes_{\phi(\alpha_2)} (H^\wedge \otimes R) \\
\downarrow & & \downarrow \\
(X^\wedge \otimes R) \otimes_{\rho_1^2} S_2 & \underset{\tau_2}{\tilde{\to}} & (X^\wedge \otimes R) \otimes_{\rho_2^2} S_2
\end{array}$$

Now, consider $\rho_1' = \rho_1 \circ \phi(\alpha_2)$. Then, put $S_3 = S/\underline{m}^3$, and consider the commutative diagram

$$\begin{array}{ccc}
(X^\wedge \otimes R) \otimes_{\rho_1'} S & \underset{\tau'}{\tilde{\to}} & (X^\wedge \otimes R) \otimes_{\rho_2} S \\
\downarrow & & \downarrow \\
(X^\wedge \otimes R) \otimes_{\rho_1'^3} S_3 & \underset{\tau_3}{\tilde{\to}} & (X^\wedge \otimes R) \otimes_{\rho_2^3} S_3 \\
\downarrow & & \downarrow \\
(X^\wedge \otimes R) \otimes_{\rho_1'^2} S_2 & = & (X^\wedge \otimes R) \otimes_{\rho_2^2} S_2
\end{array}$$

It follows from the commutativity of the lower square that $\rho_1'^{3} = \rho_2^3$, therefore that τ_3 is an automorphism.
Now, copy the procedure above, get $\alpha_3 : H^{\wedge}\otimes T^0 \to H^{\wedge}\otimes R$ such that if $\rho_1'' = \rho_1' \circ \phi(\alpha_3)$ then $\rho_1''^{4} = \rho_2^4$ etc. See now that the corresponding $\alpha_n \in \underline{\mathrm{aut}}_X(H^{\wedge}\otimes R)$ $n>2$, have exactly the properties of (ii).

Q.E.D.

There is an obvious homomorphism of group functors

$$\sigma : \underline{\mathrm{Aut}}_X \to \underline{I}_X/\underline{i}_X^*$$

In fact, to each automorphism $\alpha \in \underline{\mathrm{Aut}}_X(R) = \mathrm{Aut}_R(X\otimes R)$ there exists an automorphism $\phi(\alpha) \in \underline{I}_X(R)$ such that the following diagram commutes

$$\begin{array}{ccc} X^{\wedge}\otimes R & \overset{\sim}{\to} & (X^{\wedge}\otimes R) \underset{\phi(\alpha)}{\otimes} (H^{\wedge}\otimes R) \\ \downarrow & & \downarrow \\ X\otimes R & \underset{\alpha}{\overset{\sim}{\to}} & X\otimes R \end{array}$$

just as above. It is clear that the class of $\phi(\alpha)$ in $\underline{I}_X(R)/\underline{i}_X^*(R)$ is unique, and one checks easily that the map $\alpha \to \sigma(\alpha) =$ class of $\phi(\alpha)$ is a group homomorphism.

Let $\underline{\mathrm{Aut}}_X^1$ (resp. $\underline{\mathrm{aut}}_X^1$) be the subgroup-functor of $\underline{\mathrm{Aut}}_X$ (resp. $\underline{\mathrm{aut}}_X$) consisting for each R of those automorphisms of $X\otimes R$ that lifts to $X^{\wedge}\otimes R$.
With these notations we have the following:

<u>Corollary (2.3)</u>. (i) There is a canonical action of $I(X)/i(X)$ on $H_0^{\wedge}$.

(ii) $\underline{\mathrm{Aut}}_X/\underline{\mathrm{Aut}}_X^1 \simeq \underline{I}_X/\underline{i}_X^*$ as group-functors.

<u>Proof</u>. By (2.1) $H_0^{\wedge}$ is the maximal quotient of $H^{\wedge}$ such that for all R in ℓ, $H_0^{\wedge}\otimes R$ is a quotient of $H^{\wedge}\otimes R/\{h-ih \mid h\in H^{\wedge}\otimes R,\ i\in \underline{i}_X(R)\}$.

Let $i' \in \underline{I}_X(R)$. Since $\underline{i}_X(R)$ is normal in $\underline{I}_X(R)$, $i'(h-ih) = i'h-i'ih = (i'h)-j(i'h)$ where $i \in \underline{i}_X(R)$ and where $j \in \underline{i}_X(R)$ is defined by $i'i = ji'$. Thus $I(X) = \underline{I}_X(k)$ operates on H_0. The rest is clear.

Q.E.D.

Remark (2.4). (i) Notice that although $\underline{\text{aut}}_{\hat{X}_0}$ is smooth on $\hat{H}_0$, the group functor $\underline{\text{Aut}}_{\hat{X}_0}$ is not, in general, smooth. An example is furnished by any hyperelliptic curve X of genus ≥ 3. In this case we have $\hat{H} = \hat{H}_0$ but the involution is never liftable to $\hat{H}/\underline{m}^2$, see [La-Lø].

(ii) $I(X)/i(X)$ does not, in general, operate effectively on $\hat{H}_0$. In fact if $X = \text{Spec}(k[\underline{x}]/(f))$ where f is quasihomogeneous, the torus action $\tau \in \text{Aut}_k(X)$ is not, in general, liftable to $\hat{X}$, but τ operates trivially on $\hat{H}_0$.

(iii) The subgroup of $\text{Aut}_k(\hat{H}_0)$ consisting of those ψ for which there exists an isomorphism $\chi_\psi : \hat{X}_0 \simeq \hat{X}_0 \otimes_\psi H_0$, is a quotient of $I(X)/i(X)$. In fact let ψ be any such automorphism and consider the restriction of χ_ψ to the special fiber, $\phi: X \tilde{\to} X$. Consider further any representative $\bar{\sigma}(\phi^{-1}) \in I(X)$ of $\sigma(\phi^{-1}) \in I(X)/i(X)$. Let ψ' be the restriction of $\bar{\sigma}(\phi^{-1})$ to $\hat{H}_0$. Then there is an isomorphism $\Psi_{\psi'} : \hat{X}_0 \tilde{\to} \hat{X}_0 \hat{\otimes}_{\psi'} H_0$ and the composition

$$\hat{X}_0 \xrightarrow{\Psi_{\psi'}} \hat{X}_0 \hat{\otimes}_{\psi'} H_0 \xrightarrow{(\chi_\psi) \otimes_{\psi'} 1_{\hat{H}_0}} (\hat{X}_0 \hat{\otimes}_\psi H_0) \hat{\otimes}_{\psi'} H_0$$

reduces to the identity on the special fiber.

It follows that $\psi \circ \psi' \in \text{Aut}_k(\hat{H}_0)$ conserves the universal family $\hat{X}_0$, thus $\psi \circ \psi' = 1_{H_0}$. But then ψ is induced by $\sigma(\phi) \in I(X)/i(X)$.

(iv) The action of $I(X)$ on $\hat{H}$ induces a linear action on the tangent space $A^1(k,X;O_X)$. Since $i(X)$ acts trivially on the tangent space of $\hat{H}$, we obtain a linear action of $I(X)/i(X)$ on $A^1(k,X;O_X)$. This action is well understood and has been used in many instances, see e.g. [La-Lø]. It is, in view of (2.3) (ii), given in terms of the action

$$\textstyle\sum : \text{Aut}(X) \to G\ell(A^1(k,X;O_X))$$

defined by $\sum(\psi) = \psi^{*-1} \circ \phi_*$.

The above picture is better understood if we look at it at the Lie algebra level.

Given a functor $\underline{F}$ of groups defined on $\underline{\ell}$. Recall the definition of the Lie algebra lie $\underline{F}$, see [G-D],

$$\text{lie } \underline{F}(R) = \ker \{F(R \otimes k[\varepsilon]) \to F(R)\}.$$

Notice that

$$\text{lie } \underline{\text{aut}}_{S^\wedge}(k) = \text{Der}_k^c(S^\wedge)$$

and that by definition

$$\text{lie } \underline{i}_\pi(k) = \{D \in \text{Der}_k^c(S^\wedge) \mid Y^\wedge \otimes k[\varepsilon] \overset{\chi}{\simeq} (Y^\wedge \otimes k[\varepsilon]) \underset{\psi}{\otimes} (S^\wedge \otimes k[\varepsilon])$$
$$\text{where } \psi = \text{id} + \varepsilon \cdot D \text{ and } \chi \underset{k[\varepsilon]}{\otimes} k = \text{id}_{Y^\wedge}\}.$$

As a shorthand we shall write

$$\underline{\ell}(\pi) := \text{lie } \underline{i}_\pi, \quad \ell(\pi) := \underline{\ell}(\pi)(k)$$

This notion has the advantage, that it is readily relativized, and that it functions well with respect to functoriality. In fact, consider together with the formal family

$$\pi: Y^\wedge \to \text{Spf}(S^\wedge) = \underline{S}^\wedge$$

a morphism of complete local k-algebras

$$\rho: S^\wedge \to T^\wedge.$$

Let

$$\pi': Y'^\wedge \to \text{Spf}(T^\wedge) = \underline{T}^\wedge$$

be the pull-back, and put

$$\ell(\pi,\rho) = \{D \in \text{Der}_k^c(S^\wedge, T^\wedge) \mid (Y'^\wedge \otimes k[\varepsilon] \overset{\chi}{\simeq} (Y^\wedge \otimes k[\varepsilon]) \underset{\psi}{\otimes} (T^\wedge \otimes k[\varepsilon])$$
$$\text{where } \psi = \rho + \varepsilon \cdot D \text{ and } \chi \underset{k[\varepsilon]}{\otimes} k = \text{id}_{Y'^\wedge}\}.$$

Obviously $\ell(\pi,\rho)$ is the value at k of a functor $\underline{\ell}(\pi,\rho)$ that the reader may want to explicate. Now the restrictions of the natural morphisms

$$\text{Der}_k^c(S^\wedge, S^\wedge) \searrow \text{Der}_k^c(S^\wedge, T^\wedge) \swarrow \text{Der}_k^c(T^\wedge, T^\wedge)$$

define maps,

$$\begin{array}{ccc} \ell(\pi) & & \ell(\pi') \\ & \searrow \quad \swarrow & \\ & \ell(\pi,\rho). & \end{array}$$

Before we state the main result of this §, let us put as another shorthand

$$A^0_{\pi'} = \mathrm{im}\{A^0(T^\wedge, Y'^\wedge; O_{Y'^\wedge}) \to A^0(k,X;O_X)\}.$$

and let us recall the canonical action

$$\textstyle\sum: \underline{\mathrm{Aut}}(X) \to \underline{\mathrm{Aut}}_X/\underline{\mathrm{Aut}}^1_X \to \underline{G\ell}(A^1(k,X;O_X))$$

and the isomorphism $\underline{\mathrm{Aut}}_X/\underline{\mathrm{Aut}}^1_X \simeq \underline{I}_X/\underline{i}^*_X$, see (2.3) (ii) and (2.4) (iv) above, noticing that $A^1(k,X;O_X)\otimes_k R = A^1(R, X\otimes R, O_X\otimes R)$. One checks that lie $\underline{\mathrm{Aut}}^1_X(k) = A^0_{\pi^\wedge}$ is a Lie ideal of lie $\underline{\mathrm{Aut}}_X(k) = A^0(k,X;O_X)$ and that we therefore obtain a morphism of Lie algebras $\sigma: A^0(k,X;O_X) \to \mathrm{End}(A^1(k,X;O_X))$ that factors via

$$A^0(k,X;O_X)/A^0_{\pi^\wedge} \to (\mathrm{lie}\ \underline{I}_X/\underline{i}^*_X)(k)$$

<u>Lemma (2.5)</u>. lie $\underline{I}_X$ = lie $\underline{i}_X$.

<u>Proof</u>. Let $\delta \in \mathrm{lie}\ \underline{I}_X(R)$, then $\delta \in \mathrm{lie}\ \underline{\mathrm{aut}}_H(R)$. Since lie $\underline{\mathrm{aut}}_H(R) \simeq \mathrm{Der}_R(H^\wedge \otimes R)$, δ corresponds to a $D \in \mathrm{Der}^c_R(H^\wedge \otimes R)$ such that $\delta = \mathrm{id} + \varepsilon \cdot D \in \mathrm{aut}_{R[\varepsilon]}(H^\wedge \otimes R[\varepsilon])$. Moreover there exist isomorphisms ω and χ' such that

$$\begin{array}{ccc} X^\wedge \otimes R[\varepsilon] & \xrightarrow[\chi']{\simeq} & (X^\wedge \otimes R[\varepsilon]) \otimes_\delta (H \otimes R[\varepsilon]) \\ \uparrow & & \uparrow \\ X^\wedge \otimes R & \xrightarrow[\omega]{\simeq} & X^\wedge \otimes R \end{array}$$

commutes. Let $\chi = (\omega^{-1} \otimes \mathrm{id}_{k[\varepsilon]})\chi'$ and notice that the following diagram is commutative

$$\begin{array}{ccc} X^\wedge \otimes R[\varepsilon] & \xrightarrow[\chi]{\simeq} & (X^\wedge \otimes R[\varepsilon]) \otimes_\delta (H^\wedge \otimes R[\varepsilon]) \\ & \nwarrow \quad \nearrow & \\ & X^\wedge \otimes R & \end{array}$$

This shows that $\delta \in \mathrm{lie}\ \underline{i}_X(R)$ and we are through. Q.E.D.

From this follows that there exists an action

$$\sigma_0 : \ell(\pi^\wedge) \to \mathrm{End}(A^1(k,X;O_X))$$

which we shall use extensively.

Now let us prove the main result of this §.

<u>Theorem (2.6)</u>. (i) $\ell(\pi,\rho) = \{D \in \mathrm{Der}^c(S^\wedge,T^\wedge) \mid \exists\ E \in A^0(k,Y^\wedge,O_{Y^\wedge})$ such that for any local section y of $O_{Y^\wedge}$ and any $s \in S^\wedge$, $E(sy) = \rho(s)\cdot E(y)+D(s)\cdot y\}$

(ii) $\ell(\pi,\rho)$ is a $T^\wedge$-submodule of $\mathrm{Der}_k^c(S^\wedge,T^\wedge)$.

(ii') If $\rho = \mathrm{id}_{S^\wedge}$, then $\ell(\pi)$ is a sub k-Lie algebra of $\mathrm{Der}_c(S^\wedge)$.

(iii) There is a $T^\wedge \to k$ semilinear map

$$\lambda : \ell(\pi,\rho) \to A^0(k,X;O_X)/A^0_{\pi'}$$

which to any $D \in \ell(\pi,\rho)$ associates the class of $E \otimes_T k \in A^0(k,X;O_X)$.

(iv) Suppose $\pi = \pi^\wedge$, then λ induces an isomorphism

$$\ell(\pi^\wedge,\rho)\otimes_T k \simeq A^0(k,X;O_X)/A^0_{\pi'}.$$

(v) $H_0^\wedge$ is the maximal quotient of $H^\wedge$ trivializing $\ell(\pi^\wedge)$. Moreover $\ell(\pi^\wedge)\otimes_{H^\wedge} H_0^\wedge$ is an $H_0^\wedge$-Lie algebra, and a deformation of the k-Lie algebra $L(X) = A^0(k,X;O_X)/A^0_\pi$.

<u>Proof</u>. To simplify notations put $S = S^\wedge$, $T = T^\wedge$, $H = H^\wedge$ and $Y = Y^\wedge$. Let $D \in \ell(\pi,\rho)$, then $\phi = \rho+\varepsilon\cdot D$ is a morphism of complete local k-algebras.

$$\phi : S \to T\otimes k[\varepsilon] = T[\varepsilon]$$

is such that there exists an $T\otimes k[\varepsilon]$-isomorphism

$$\chi_D : Y\otimes_S T[\varepsilon] \tilde{\to} Y\otimes_\phi T[\varepsilon]$$

lifting the identity on $Y \otimes_S T$.

There are commutative diagrams of morphisms of k-schemes,

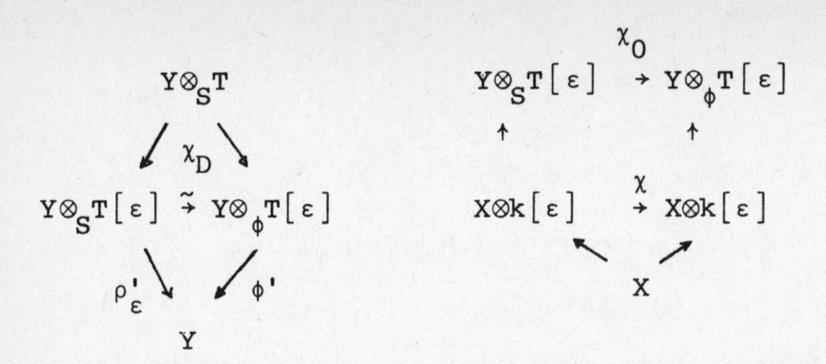

The difference $E' = \phi' \circ \chi_D - \rho'_\varepsilon$ corresponds to an element E of $A^0(k,Y,O_Y \otimes_S T)$. If y is a local section of O_Y and $s \in S$, then $\rho'_{\varepsilon *}$ maps sy to $\rho(s)y$ in $O_Y \otimes_S T[\varepsilon]$, and ϕ'_* maps sy to $(\rho(s)+\varepsilon \cdot D(s))y$ in $O_Y \otimes_\phi T[\varepsilon]$. Since χ_D is the identity on $Y \underset{S}{\otimes} T$, $\chi_{D*}((\rho(s)+\varepsilon D(s))y) = (\rho(s)+\varepsilon D(s))(y+\varepsilon E(y))$. Thus $E'_*(sy) = \varepsilon(\rho(s)E(y)+D(s)y)$ and consequently

$(*)$ $E(sy) = \rho(s)E(y) + D(s)y$.

If on the other hand $E \in A^0(k,Y,O_Y \otimes_S T)$ is such that $(*)$ holds, for some derivation $D \in \mathrm{Der}_k(S,T)$ then for $\phi = \rho + \varepsilon D$ and $\chi_{D*}(y \otimes 1) = y + \varepsilon E(y)$, we find that χ_D is an isomorphism between $Y \otimes_S T[\varepsilon]$ and $Y \otimes_\phi T[\varepsilon]$ lifting the identity on $Y \otimes_S T$, therefore $D \in \ell(\pi,\rho)$, and we have proved (i). Since $A^0(k,Y;O_Y \otimes_S T)$ is a T-module (ii) follows. If $E_i \in A^0(k,Y,O_Y)$ correspond to D_i as in (i), for $i = 1,2$, then one checks that

$$[E_1,E_2](sy) = s[E_1,E_2](y)+[D_1,D_2](s) \cdot y$$

therefore $\ell(\pi)$ is a sub Lie algebra of $\mathrm{Der}_k(S,S)$ and (ii') follows. (iii) is a consequence of (i), as $\chi_D \otimes_T k = \chi = \mathrm{id} + \varepsilon\lambda(D)$, compare the right hand diagram above. Now, for the proof of (iv), we first notice that λ is surjective. In fact if $\bar{E} \in A^0(k,X;O_X)$, then $\chi = \mathrm{id}_X + \varepsilon \bar{E} \in \mathrm{Aut}_{k[\varepsilon]}(X \otimes k[\varepsilon])$. As above (see (2.3) (ii)), there is a $\bar{\phi}: H \to H \otimes k[\varepsilon]$ such that $X^\wedge \otimes k[\varepsilon]$ and $X^\wedge \otimes_{\bar{\phi}} H[\varepsilon]$ are isomorphic, the isomorphism lifting $\mathrm{id}_X + \varepsilon \bar{E}$. Tensorise with T and obtain $\phi: H \to T[\varepsilon]$ such that $X^\wedge \otimes_H T[\varepsilon]$ and $X^\wedge \otimes_\phi T[\varepsilon]$ are isomorphic, the isomorphism lifting $id_X+\varepsilon\bar{E}$. Since $id_X+\varepsilon\bar{E}\in\ker(Aut_{k[\varepsilon]}(X\otimes k[\varepsilon]) \to Aut_k(X))$ we may choose $\bar{\phi}$ such that $\bar{\phi} = id+\varepsilon\bar{D}$, with $\bar{D}\in Der_k(H)$. In fact the composition $\theta:H \xrightarrow{\bar{\phi}} H[\varepsilon] \to H$ is such that $X^\wedge\otimes_\theta H$ and $X^\wedge$ are isomorphic, the isomorphism lifting the identity on X. Then θ is necessarily an automorphism. Consider the composition $\bar{\phi}^1:H \xrightarrow{\theta^{-1}} H \xrightarrow{\bar{\phi}} H[\varepsilon]$ and see that $\bar{\phi}^1$ has the required property. But then $\phi = \rho+\varepsilon\cdot D$, with $D\in\ell(\pi,\rho)$, and we have proved that λ is surjective.

To complete the proof of (iv) we shall prove that for any basis $\{\bar{E}_i\}_{i=1}^N$ of A^0/A^0_{π}, with $\bar{E}_i = \lambda(D_i)$, and $D_i\in\ell(\pi^\wedge,\rho)$ corresponding to $E_i\in A^0(k,X^\wedge;O_X{}^\wedge\otimes_H T))$, the D_i's generate $\ell(\pi^\wedge,\rho)$ as a T-module.

Pick any such basis $\{\bar{E}_i\}_{i=1}^N$ and corresponding D_i's, and E_i's. Let $h_i\in T$, then $\sum h_iD_i\in\ell(\pi^\wedge,\rho)$ corresponds to $\sum_i h_iE_i$. Consequently we need only prove that for any $D\in\ell(\pi^\wedge,\rho)$, the corresponding E is a sum of the form $\sum_i h_iE_i$, modulo $A^0(H,X^\wedge;O_X{}^\wedge\otimes_H T)$. But this is easily achieved. In fact, the image of E in $A^0(k,X,O_X)$ can be written as $\sum_{i=1}^N h_i^0\bar{E}_i + \bar{E}(0)$, $\bar{E}(0)\in A^0_{\pi}$.

Let $E(0)\in A^0(H,X^\wedge;O_X{}^\wedge\otimes_H T)$ be the preimage of $\bar{E}(0)$, then D corresponds to $E-E(0)$ as well, and $\bar{E}-\bar{E}(0) = \sum_{i=1}^N h_i^0\bar{E}_i$. Therefore $E-E(0) -\sum_{i=1}^N h_i^0E_i$ maps to zero in $A^0(k,X,O_X)$, hence also in $A^0(k,X^\wedge;O_X)$. Notice that this is a consequence of the fact that D and the D_i's map the maximal ideal $\underline{m} \subseteq H$ into the maximal ideal $\underline{n} \subseteq \underline{T}$.
Consider the exact sequence

$$o \to A^0(k,X^\wedge;O_X{}^\wedge\otimes_H \underline{n}/\underline{n}^2) \to A^0(k,X^\wedge;O_X{}^\wedge\otimes_H T/\underline{n}^2) \to A^0(k,X^\wedge;O_X{}^\wedge\otimes_H k)$$

Obviously:

$$A^0(k,X^\wedge;O_X{}^\wedge\otimes_H \underline{n}/\underline{n}^2) = A^0(k,X^\wedge;O_X)\otimes\underline{n}/\underline{n}^2$$

$$A^0(k,X^\wedge;O_X{}^\wedge\otimes_H k) = A^0(k,X^\wedge;O_X).$$

Therefore the image of $(E-E(0)-\sum_{i=1}^N h_i^0 E_i)$ in $A^0(k,X^\wedge;O_X\wedge\otimes T/\underline{n}^2)$ sits in $A^0(k,X^\wedge;O_X\wedge\otimes_H \underline{n}/\underline{n}^2)$ and in fact in the sub vector space $A^0(k,X;O_X)\otimes\underline{n}/\underline{n}^2$. This follows since $E-E(0)-\sum_{i=1}^N h_i^0 E_i$ corresponds to $D-\sum_{i=1}^N h_i^0 D_i$. By construction $\lambda(D-\sum_{i=1}^N h_i^0 D_i)=0$ such that by (2.9), $(D-\sum_{i=1}^N h_i^0 D_i)(\underline{m})\subseteq \underline{n}^2$. Therefore there must exist $h_i^1\in\underline{n}\subseteq T$ such that the image of $(E-E(0)-\Sigma h_i^0 E_i)-\Sigma h_i^1 E_i$ in $A^0(k,X^\wedge;O_X\wedge\otimes_H T/\underline{n}^2)$ sits in $A^0_{\pi'}\otimes\underline{n}/\underline{n}^2$. Thus there exists $E(1)\in\underline{n}\cdot A^0(H,X^\wedge;O_X\wedge\otimes_H T)$ such that $E-(E(0)+E(1))-\sum_{i=1}^N(h_i^0+h_i^1)E_i$ maps to zero in $A^0(k,X^\wedge;O_X\wedge\otimes_H T/\underline{n}^2)$. Continue, considering the exact sequences induced by $0\to\underline{n}^p/\underline{n}^{p+1}\to T/\underline{n}^{p+1}\to T/\underline{n}^p\to 0$, we obtain $E=\sum_{i=1}^N h_i E_i+\sum_{j=0}^\infty E(j)$ where $h_i=\sum_{j=0}^\infty h_i^j\in T$, $E(j)\in A^0(H,X^\wedge;O_X\wedge\otimes_H T)$ and $\sum_{j=0}^\infty E(j)$ converges to an element of $A^0(H,X^\wedge;O_X\wedge_H T)$.

Now, to prove (v), notice that it follows from (iv), with $\rho:H^\wedge\to H_0^\wedge$, and from (1.6) that $H_0^\wedge$ is the maximal quotient of $H^\wedge$ trivializing $\ell(\pi^\wedge)$. Therefore if $H_0^\wedge=H^\wedge/\mathfrak{a}$ any $\delta\in\ell(\pi^\wedge)$ maps $H^\wedge$ into $\mathfrak{a}$. Moreover, if $k\in H$, $\delta_1,\delta_2\in\ell(\pi^\wedge)$ then $[\delta_1,k\delta_2]=k[\delta_1,\delta_2]+\delta_1(k)\delta_2$. Consequently $\mathfrak{a}\cdot\ell(\pi^\wedge)$ is a Lie ideal of $\ell(\pi^\wedge)$ and $\ell(\pi^\wedge)\otimes_{H^\wedge}H_0^\wedge$ is an $H_0^\wedge$-Lie algebra. Q.E.D.

<u>Remark (2.7)</u>. In particular we have proved that if $\dim_k(A^0/A^0_\pi\wedge)<\infty$, then $\ell(\pi^\wedge)$ is of finite type as $H^\wedge$-module and vice versa. Moreover the rank of $\ell(\pi^\wedge)$ is bounded by $\dim_k(A^0/A^0_\pi\wedge)$.

For every $D\in\ell(\pi,\rho)$, D is a derivation of $H^\wedge$ into $T^\wedge$ mapping the maximal ideal $\underline{m}$ into $\underline{n}$. Therefore D induces an homomorphism D_* of the tangent space of $T^\wedge$, into the tangent space of $H^\wedge$. Because of (2.6) (iv) (in fact (2.9)) the map $D\to D_*$,

$$\sigma_0:\ \ell(\pi^\wedge,\rho)\to \mathrm{Hom}_k(\underline{m}/\underline{m}^2,\ \underline{n}/\underline{n}^2)=\mathrm{Hom}_k((\underline{n}/\underline{n}^2)^*,\ A^1(k,X;O_X))$$

factors via $A^0(k,X;O_X)/A^0_{\pi'}$. In particular, we find a homomorphism, of Lie algebras

$$\rho:A^0(k,X;O_X)\to \mathrm{End}(A^1(k,X;O_X)),$$

which we shall make explicit. Put for $\bar{E}\in A^0(k,X;O_X)$, $\bar{E}_*=\rho(\bar{E})$.

Notice first that any non-zero element $\xi \in A^1(k,X;O_X)$ may be considered as a surjective morphism of complete local k-algebras

$$\xi: H^\wedge \to k[\eta], \quad \eta^2 = 0.$$

Consider

$$\ell(\pi^\wedge,\xi) = \{D \in \mathrm{Der}_*(H^\wedge,k[\eta]) \mid X^\wedge \underset{\xi}{\otimes} k[\eta] \otimes k[\varepsilon] \overset{\chi}{\to} (X^\wedge \otimes k[\varepsilon]) \otimes_\psi k[\eta,\varepsilon],$$

$$\psi = \mathrm{id} + \varepsilon \cdot D, \quad \chi \otimes_{k[\varepsilon]} k = \mathrm{id}_{X^\wedge \otimes k[\eta]}\}.$$

By (2.2) (ii) there exists an element $\alpha \in \underline{i}_X(k[\varepsilon])$ such that $(\xi \otimes \mathrm{id}_{k[\varepsilon]}) \circ \alpha = \xi + \varepsilon \cdot D$. Since $\alpha = \mathrm{id} + \varepsilon \cdot E$, $E \in \mathrm{lie}\ \underline{i}_X$ we obtain $D = \xi \circ E$. On the other hand if $E \in \mathrm{lie}\ \underline{i}_X$ then certainly $\xi \circ E \in \ell(\pi^\wedge,\xi)$. Since $\eta \cdot \mathrm{Der}_*(H^\wedge,k[\eta]) = 0$, (2.6) (iv) implies the following

<u>Corollary (2.8)</u>. The orbit of ξ under $\ell(\pi^\wedge)$ is isomorphic to

$$\ell(\pi^\wedge,\xi) \simeq A^0(k,X;O_X)/A^0_\xi$$

where A^0_ξ is the subspace of $A^0(k,X;O_X)$ of those elements that lift to the family X_ξ on $k[\eta]$.

Notice that the proof of the following result is independent upon (2.6) (iv).

<u>Proposition (2.9)</u>. Let $D \in \ell(\pi^\wedge,\rho)$ and let $\bar{E} \in A^0(k,X;O_X)$ represent $\lambda(D) \in A^0(k,X;O_X)/A^0_\pi$, then $\sigma_0(D) \in \mathrm{Hom}_k((\underline{n}/\underline{n}^2)^*, A^1(k,X;O_X))$ is defined in the following way:

Given $\xi \in (\underline{n}/\underline{n}^2)^*$, let X_ξ be the corresponding lifting of X to $k[\eta]$, defined by the composition $H^\wedge \to T^\wedge \to T^\wedge/\underline{n}^2 \overset{\xi}{\to} k[\eta]$ and consider the lifting situation; $\varepsilon^2 = \eta^2 = 0$

$$\begin{array}{ccccc}
\mathrm{Spec}(k[\varepsilon][\eta]) & \to X_\xi \otimes k[\varepsilon] & \dashrightarrow & X_\xi \otimes k[\varepsilon] \\
\uparrow & \uparrow & & \uparrow \\
\mathrm{Spec}(k[\varepsilon]) & \to X \otimes k[\varepsilon] & \xrightarrow[\mathrm{id}+\varepsilon\cdot\bar{E}]{} & X \otimes k[\varepsilon] \\
\uparrow & \uparrow & & \uparrow \\
\mathrm{Spec}(k) & \to X & = & X
\end{array}$$

Let $o(\mathrm{id}+\varepsilon\bar{E})$ be the obstruction for lifting the automorphism $\mathrm{id}+\varepsilon\bar{E}$ as an automorphism of $X_\xi \otimes k[\varepsilon]$, then $o(\mathrm{id}+\varepsilon\bar{E}) \in A^1(k[\varepsilon], X \otimes k[\varepsilon]; O_X \otimes k[\varepsilon]) = A^1(k,X;O_X) \otimes k[\varepsilon]$, has the form $\varepsilon \cdot \bar{E}_*(\xi)$.

<u>Proof</u>. Let $D \in \ell(\pi,\rho)$ correspond to $E \in A^0(k,X^\wedge;O_{X^\wedge}\otimes_{H^\wedge}T^\wedge)$ as in (2.6) (i) and consider $\bar{E}\in A^0(k,X;O_X)$, $\bar{E}=E\otimes_{H^\wedge}k$. $\sigma_0(D)\in \mathrm{Hom}(A^{1*}, \underline{n}/\underline{n}^2)$ corresponds to the morphism $\delta=\rho+\varepsilon\cdot D\in \mathrm{Mor}_{k[\varepsilon]}(H^\wedge\otimes k[\varepsilon],\ T^\wedge\otimes k[\varepsilon])$. By construction, there exists an isomorphism χ_δ such that the following diagram commutes

$$
\begin{array}{ccc}
X^\wedge\otimes k[\varepsilon] & \xleftarrow{\chi_\delta} & (X^\wedge\otimes k[\varepsilon]) \underset{\delta}{\otimes} (T^\wedge\otimes k[\varepsilon]) \\
\uparrow & & \uparrow \\
X\otimes k[\varepsilon] & \xleftarrow[\mathrm{id}+\varepsilon\cdot\bar{E}]{} & X\otimes k[\varepsilon]
\end{array}
$$

Reduce the diagram modulo $\underline{m}_H^2\wedge$ and obtain

$$
\begin{array}{ccc}
X_2^\wedge\otimes k[\varepsilon] & \xleftarrow{\chi_{\delta_2}} & (X_2^\wedge\otimes k[\varepsilon])\otimes_{\delta_2}(T_2\otimes k[\varepsilon]) \\
\uparrow & & \uparrow \\
X\otimes k[\varepsilon] & \xleftarrow[\mathrm{id}+\varepsilon\bar{E}]{} & X\otimes k[\varepsilon]
\end{array}
$$

where $\delta_2=\rho_2+\varepsilon\cdot D_2$ and D_2 is the linearization of D, i.e. the derivation of $\rho_2:H_2=H^\wedge/\underline{m}^2\to T_2=T^\wedge/\underline{n}^2$ induced by D. Notice that since ξ is a linear form on $\underline{n}/\underline{n}^2$ inducing a k-algebra homomorphism $\xi:T_2\to k[\eta]$ and a composition $\bar{\xi}:H_2\to T_2\to k[\eta]$, ξ corresponds to a lifting X_ξ of X to $k[\eta]$, thus to an element $\xi\in A^1(k,X;O_X)$. Consider the homomorphism $\xi\otimes 1_{k[\varepsilon]}:T_2\otimes k[\varepsilon]\to k[\eta]\otimes k[\varepsilon]$ and the induced commutative diagram

$$
\begin{array}{ccc}
(X_2^\wedge\otimes k[\varepsilon]) \underset{\bar{\xi}\otimes 1_{k[\varepsilon]}}{\otimes} (k[\eta]\otimes k[\varepsilon]) & \xleftarrow{\chi_{\delta_2}\otimes(\bar{\xi}\otimes 1_{k[\varepsilon]})} & (X_2^\wedge\otimes k[\varepsilon])\otimes_{\delta_2}(T_2\otimes k[\varepsilon]) \underset{\xi\otimes 1_{k[\varepsilon]}}{\otimes} (k[\eta]\otimes k[\varepsilon]) \\
\uparrow & & \uparrow \\
X\otimes k[\varepsilon] & \xleftarrow[\mathrm{id}+\varepsilon\cdot\bar{E}]{} & X\otimes k[\varepsilon]
\end{array}
\qquad (*)
$$

where $(X_2^\wedge\otimes k[\varepsilon]) \underset{\bar{\xi}\otimes 1_{k[\varepsilon]}}{\otimes} (k[\eta]\otimes k[\varepsilon]) = X_\xi\otimes k[\varepsilon]$.

The composition $(\xi\otimes 1_{k[\varepsilon]})\circ\delta_2:H_2\otimes k[\varepsilon]\to k(\eta)\otimes k[\varepsilon]$ is by definition of $\sigma_0(D)$, $\bar{\xi}\otimes 1_{k[\varepsilon]}+\varepsilon(\sigma_0(D)(\xi)\otimes 1_{k[\varepsilon]})$ where $\sigma_0(D)(\xi)\in A^1(k,X;O_X)$ is considered as an algebra homomorphism $\sigma_0(D)(\xi):H_2\to k[\eta]$, as explained above. But then $(X_2^\wedge\otimes k[\varepsilon])\otimes_{\delta_2}(T_2\otimes k[\varepsilon]) \underset{\xi\otimes 1_{k[\varepsilon]}}{\otimes} (k[\eta]\otimes k[\varepsilon])$ is the lifting of $X\otimes k[\varepsilon]$ corresponding to $\xi\cdot\eta+\sigma_0(D)(\xi)\cdot\eta\cdot\varepsilon\in A^1(k[\varepsilon], X:O_X\otimes k[\varepsilon]\otimes(\eta))$. From the existence of the commutative diagram $(*)$ we deduce

$$(id+\varepsilon\cdot\bar{E})^{*-1}(id+\varepsilon\bar{E})_*(\xi\otimes 1_{k[\varepsilon]}) = \xi\otimes 1_{k[\varepsilon]}+\varepsilon(\sigma_0(D)\otimes 1_{k[\varepsilon]})$$

where $(id+\varepsilon\bar{E})^*: A^1(k[\varepsilon],X\otimes k[\varepsilon],O_X\otimes k[\varepsilon]) \to A^1(k[\varepsilon],X\otimes k[\varepsilon],O_X\otimes k[\varepsilon])$ is the automorphism induced by $id+\varepsilon\bar{E}: X\otimes k[\varepsilon] \to X\otimes k[\varepsilon]$ and $(id+\varepsilon\bar{E})_*$ is the one induced by $id+\varepsilon\bar{E}: O_X\otimes k[\varepsilon] \to O_X\otimes k[\varepsilon]$. But then $\varepsilon\cdot\sigma_0(D)(\xi)$ is the obstruction for lifting $id+\varepsilon\bar{E}$ to $X_\xi\otimes k[\varepsilon]$.

Q.E.D.

Corollary (2.10). Let $\bar{E} \in A^0(k,X;O_X)$ then $\bar{E}_* \in End(A^1)$ is defined by

$$\varepsilon\cdot\bar{E}_* = (id+\varepsilon\bar{E})_*(id+\varepsilon\cdot\bar{E})^{*-1}-id$$

Proof. This is exactly the contention of the proof above. Q.E.D.

Corollary (2.11). Let $X = Spec(k[x_1,\dots,x_n]/J)$ and let $D \in Der_k(k[\underline{x}]/J)$. Then the action of $\rho(D)$ on

$$H^1(k,X,O_X) = Hom_{k[\underline{x}]}(J,k[\underline{x}]/J)/Der$$

is defined as follows. Let $\xi \in H^1(k,X;O_X)$ be represented by a homomorphism $\bar{\xi}:J \to k[\underline{x}]/J$ and lift D to a derivation $E:k[\underline{x}] \to k[\underline{x}]$ then $\rho(D)(\xi)$ is represented by the homomorphism

$$D\circ\bar{\xi}-\bar{\xi}\circ E:J \to k[\underline{x}]/J.$$

Proof. Use (2.10). Q.E.D.

In particular we have:

Corollary (2.12). Let $X = Spec(k[\underline{x}]/(f))$ and let $\bar{E}\in Der_k(k[\underline{x}]/(f))$ be defined by $\sum_{i=1}^N \frac{\partial f}{\partial x_i}\bar{E}(x_i) = q(\underline{x})f(\underline{x})$. Then if ξ is the class of the polynomial $\xi(\underline{x})$ in $H^1(k,k[\underline{x}]/(f),k[\underline{x}]/(f)) =$ $k[\underline{x}]/(f,\frac{\partial f}{\partial x_1},\dots,\frac{\partial f}{\partial x_n})$, the action of $\bar{E}_*$ on ξ is given by

$$\bar{E}_*(\xi) = \text{class of } (\sum_{i=1}^n \frac{\partial\xi(\underline{x})}{\partial x_i}\bar{E}(x_i)-q(\underline{x})\xi(\underline{x}))$$

<u>Corollary (2.13)</u>. The rank of the linearized action of $\ell(\pi)$ on $\hat{H}$ is equal to the dimension of the maximal orbit of $A^1(k,X;O_X)$ under the action of $A^0(k,X;O_X)$.

<u>Proof</u>. By definition, the linearized action of $\ell(\pi)$ on $\hat{H}$ is the one given in terms of the action of $\ell(\pi)$ on the tangent space $(\underline{m}/\underline{m}^2)^* = A^1(k,X;O_X)$. The rank of $\ell(\pi)$ is then almost by definition the dimension of the maximal orbit of $A^1(X,O_X)$ under $\ell(\pi)$.

Q.E.D.

§3. THE KODAIRA-SPENCER MAP AND ITS KERNEL

Introduction. In this § we shall apply the results of §1 and §2 to study the local properties of formally versal families of objects of the type we are concerned with, see (3.6) for the definition of formal versality, and notice that the families we are talking about are algebraic families, not formal ones.

We start by defining the Kodaira-Spencer map $g: \mathrm{Der}_k(S) \to A^1(S,Y;O_Y)$ associated to a flat family $\eta: Y \to \underline{S}$.

The first part of the § is a study of the properties of the kernel V_η of g. As we shall see V_η is a sub k-Lie algebra of $\mathrm{Der}_k(S)$ with nice functorial properties.

We shall not venture into the difficult problem of when formally versal algebraic families exist. At this point, we take the easy way out, assuming that the objects X we are handling are such that

(A_1) there exists an algebraization $\pi: \tilde{X} \to \underline{H} = \mathrm{Spec}(H)$ of the formal versal family $\hat{\pi}: \hat{X} \to \hat{\underline{H}}$

and such that

(V) π is formally versal.

We then formulate a set of conditions (V'), akin to the conditions studied by M. Artin, see [Ar], which imply (V) . Assuming (V') we prove that the infinitesimal notions of §1 and §2, such as the prorepresenting substratum $\hat{\underline{H}}_0$ and the Lie algebra $\ell(\hat{\pi})$ are formalizations of local notions, $\underline{H}_0$ and V respectively, see (3.5).

Since the main objective of this § is the construction of a local moduli suite, we shall have to assume that π is versal in the étale topology, i.e.

(A_2) If $\eta: Y \to \underline{S}$ is a flat family with fiber $X = \eta^{-1}(\underline{s})$ for some closed point $\underline{s} \in \underline{S}$, then there exists an étale neighbourhood $\varepsilon: \underline{E} \to \underline{S}$ of $\underline{s}$ and a morphism $\rho: \underline{E} \to \underline{H}$ such that there is an isomorphism $\varepsilon^*(\eta) \simeq \rho^*(\pi)$ between the pull-backs of η and π.

We know, see [E], that (A_1) holds for affine schemes with isolated singularities. In [Gr 1], Théoréme (5.4.5), Grothendieck gives conditions for when (A_1) holds for projective schemes.

Moreover, (A_2) holds, see [Ar], when X is projective, when X is a finitely generated graded k-algebra and we restrict to deformations within the category of graded k-algebras, or when X is a complete local k-algebra of finite type, restricting this time the deformations to the category of such formal, or algebroid, schemes. This last example is going to be treated in some detail in §4-§6. Notice that (A_2) does not hold in the affine case, see the main introduction to this paper.

Assume that (A_1), (A_2) and (V') hold. Then the flattening stratification $\{\underline{S}_\tau\}$ of $\underline{H}$ corresponding to the H-module $V=V_\pi$, is stable under the action of V, and the main results of this paragraph may now be summed up as follows: There exists, see (3.18), a quotient $\underline{M}_\tau$ of $\underline{S}_\tau$, in the category of algebraic spaces, and a cartesian square:

$$\begin{array}{ccc} \tilde{X}_\tau & \rightarrow & \overline{X}_\tau \\ \pi_\tau \downarrow & \square & \downarrow \overline{\pi}_\tau \\ \underline{S}_\tau & \rightarrow & \underline{M}_\tau \end{array}$$

where π_τ is the restriction of π to $\underline{S}_\tau$, such that if $\overline{\pi}_\nu$ is constant on a subscheme Y of $\underline{M}_\tau$, then Y is finite and reduced.

Put $X(\underline{t}) = \pi^{-1}(\underline{t})$ for a closed point $\underline{t}$ of $\underline{H}$. Then the closed points of $\underline{S}_\tau$ are those $\underline{t} \in \underline{H}$ for which $\dim_k A^1(k,X(\underline{t});O_{X(\underline{t})}) = \tau$, and we may rephrase the property of $\overline{\pi}_\tau$ above in the following way: $\underline{M}_\tau$ prorepresents the τ-constant deformations of X in the family π_τ.

The collection $\{\underline{M}_\tau\}$ is by definition (3.25) <u>the local moduli suite of X</u>.

Strictly speaking $\underline{M}_\tau$ is not, in general, a subset of the set of isomorphism classes of deformations of X. We are therefore, in (3.25) slightly abusing the language used in the main introduction.

Notice that we do not claim that each $\underline{M}_\tau$ is neither a fine moduli nor a coarse moduli space.

Of more interest for the applications we have in mind, we prove, see (3.24), that for every component $\underline{S}_{\tau,c}$ of $\underline{S}_\tau$ there is an open dense subscheme of the reduced normalization $\underline{S}'_{\tau,c}$ of $\underline{S}_{\tau,c}$ on which the action of the kernel of the corresponding Kodaira-Spencer map has a good quotient in the category of schemes, whith properties like those of $\underline{M}_{\tau}$.

Notice that we do not know whether V acts rationally or not on $\underline{H}$, therefore we cannot invoke the general results of Rosenlicht, see [R], or Dixmier and Raynaud, see [D-R].

Assuming (V') we prove, see (3.12), that $V \underset{H}{\otimes} H_0$ is a deformation of the Lie algebra $L(X) = A^0(k,X;O_X)/A^0_{\pi}$. In fact, see (3.17), we obtain for every τ a flat family of Lie algebras defined on $\underline{S}_{\tau}$, the fibers of which are the k-Lie algebras $L(X(\underline{t}))$, $\underline{t} \in \underline{S}_{\tau}$.

Thus we are lead to the study of deformations of Lie algebras. Now, we might have included Lie algebras among our objects of study from the beginning of. For reasons to be explained elsewhere, we shall treat it apart.

In the above process we also find a necessary condition, and a sufficient condition for an object satisfying (A_1), (A_2) and (V) to have a rigidification, see (3.11). Moreover, we prove a result which was conjectured by Wahl [W], and recently also proved by Greuel and Loojinga [G-L], on the dimension of a smoothing component of $\underline{H}$, see (3.10).

Let S be any k-algebra. Put $\underline{S} = \mathrm{Spec}(S)$ and consider a flat family

$$\eta: Y \to \underline{S}.$$

Corresponding to the simplicial k-algebra

$$S \overset{\nu'_1}{\underset{\nu'_2}{\rightleftarrows}} S\otimes S \;\rightleftarrows\; S\otimes S\otimes S \cdots , \qquad \nu'_1 = id\otimes 1,\ \nu'_2 = 1\otimes id$$

one may define a series of obstructions for descent of Y to k.

The first descent obstruction is gotten in the following way.

Put $I = \ker\{S\otimes S \overset{m}{\to} S\}$, where m is the multiplication, and consider the diagram

$$o \leftarrow S \overset{\nu_1}{\underset{\nu_2}{\rightleftarrows}} S\otimes S/I^2 \leftarrow I/I^2 \leftarrow o , \quad I/I^2 \simeq \Omega_{S/k}.$$

Since $\nu_1^*(Y)$ and $\nu_2^*(Y)$ are two liftings of Y to $S\otimes S/I^2$, the difference

$$g(Y) = \nu_1^*(Y) - \nu_2^*(Y)$$

sits in $A^1(S,Y;\, O_Y \otimes_S \Omega_{S/k})$. $g(Y)$ is the obstruction for lifting the identity morphism on Y to a morphism between $\nu_1^*(Y) = Y \otimes_{\nu_1} (S\otimes S/I^2)$ and $\nu_2^*(Y) = Y \otimes_{\nu_2} (S\otimes S/I^2)$. Since for every S-module M, $\mathrm{Der}_k(S,M) = \mathrm{Hom}_S(\Omega_{S/k}, M)$, any $D \in \mathrm{Der}_k(S,M)$ induces a homomorphism $\partial_D : A^1(S,Y;O_Y \otimes_S \Omega_{S/k}) \to A^1(S,Y;O_Y \otimes_S M)$. Consider the S-module homomorphism

$$g_M : \mathrm{Der}_k(S,M) \to A^1(S,Y;O_Y \otimes_S M)$$

defined by

$$g_M(D) = \partial_D(g(Y)).$$

<u>Definition (3.1)</u>. The morphism g_S denoted by,

$$g_\eta :\ \mathrm{Der}_k(S) \to A^1(S,Y;O_Y)$$

is called the Kodaira-Spencer map associated to the family η.

<u>Proposition (3.2)</u>. Let $\underline{s} \in \underline{S} = \mathrm{Spec}(S)$ be a k-point. Then the induced map

$$g_\eta \otimes_S \underline{k}(\underline{s}) :\ \theta_{\underline{S},\underline{s}} \to A^1(k(\underline{s}), Y(\underline{s}); O_{Y(\underline{s})})$$

where $Y(\underline{s}) = \eta^{-1}(\underline{s})$, is the composition of the tangent map $g_{\underline{s}} : t_{\underline{S},\underline{s}} \to A^1(k(\underline{s}), Y(\underline{s}), O_{Y(\underline{s})})$ of the canonical morphism $\hat{\underline{S}}_{\underline{s}} \to \hat{\underline{H}}(\underline{s})$ defined by the formal deformation $Y \otimes_S \hat{S}_{\underline{s}}$ of $Y(\underline{s})$ to $\hat{S}_{\underline{s}}$ and the canonical map $\theta_{\underline{S},\underline{s}} \to t_{S,\underline{s}}$. (Here $\hat{H}(\underline{s})$ is the formal moduli of $Y(\underline{s})$.)

<u>Proof</u>. Let P_i, $i = 1,2$ be the ring $S \otimes_k S/I^2$ considered as left S-algebra via ν_1 and ν_2, respectively, and as right S-algebra via ν_2. Then, as left S-algebra we have the following isomorphisms

$$P_1 \otimes_S \underline{k}(s) \simeq S/\underline{m}_s^2$$

$$P_2 \otimes_S \underline{k}(s) \simeq S/\underline{m}_s^2 = \underline{k}(s)[\underline{m}_s/\underline{m}_s^2]$$

By definition, we have

$$\nu_i^*(Y) = Y \otimes_S P_i \ , \quad i = 1,2$$

therefore,

$$\nu_i^*(Y) \otimes_S \underline{k}(\underline{s}) = \begin{cases} Y \otimes S/\underline{m}_{\underline{s}}^2 & i = 1 \\ Y(s) \underset{k(\underline{s})}{\otimes} k(\underline{s})[\underline{m}_{\underline{s}}/\underline{m}_{\underline{s}}^2] & i = 2 \end{cases}$$

Consider the exact sequence of right S-modules

$$0 \leftarrow S \leftarrow P_i \leftarrow \Omega_{S/k} \leftarrow 0$$

and the commutative diagram:

$$\begin{array}{ccccc} Y & \overset{\bar{\nu}_i}{\leftarrow} & Y \otimes_S P_i & \leftarrow & Y \otimes_S P_i \otimes_S k(\underline{s}) \\ \downarrow & & \downarrow & & \downarrow \\ \mathrm{Spec}(S) & \overset{\nu_i}{\leftarrow} & \mathrm{Spec}(P_i) & \leftarrow & \mathrm{Spec}(P_i \otimes_S k(\underline{s})) = \mathrm{Spec}(S/\underline{m}^2) \end{array}$$

Obviously $g(Y) = Y \otimes_S P_1 - Y \otimes_S P_2 \in A^1(S,Y;O_Y \otimes_S \Omega_{S/k})$ is, under the specialization map $S \to k(\underline{s})$, mapped to $Y \otimes_S S/\underline{m}_{\underline{s}}^2 - Y(\underline{s}) \otimes_{k(\underline{s})} S/\underline{m}_{\underline{s}}^2 \in A^1(k(\underline{s}),Y(\underline{s});O_{Y(\underline{s})} \otimes_{k(\underline{s})} \underline{m}_{\underline{s}}/\underline{m}_{\underline{s}}^2)$.

Therefore $g \otimes k(\underline{s})$ is induced by the map

$$g_{\underline{s}} : (\underline{m}_{\underline{s}}/\underline{m}_{\underline{s}}^2)^* \to A^1(k(\underline{s}),X(\underline{s}),O_{X(\underline{s})})$$

defind by $\tilde{X} \otimes_S S/\underline{m}_{\underline{s}}^2$. Q.E.D.

We shall need a relativized version of the Kodaira-Spencer map. Let $\rho: S \to T$ be a morphism of k-algebras of finite type, and let $\eta: Y \to \underline{S}$ be a flat morphism of finite type, then for $Y' = Y \times_{\underline{S}} \underline{T} = Y \otimes_S T$ $\eta': Y' \to \underline{T}$ is flat. Putting $S[\varepsilon] = S \otimes k[\varepsilon]$, $\rho' = \mathrm{Spec}(\rho)$, we obtain a cube:

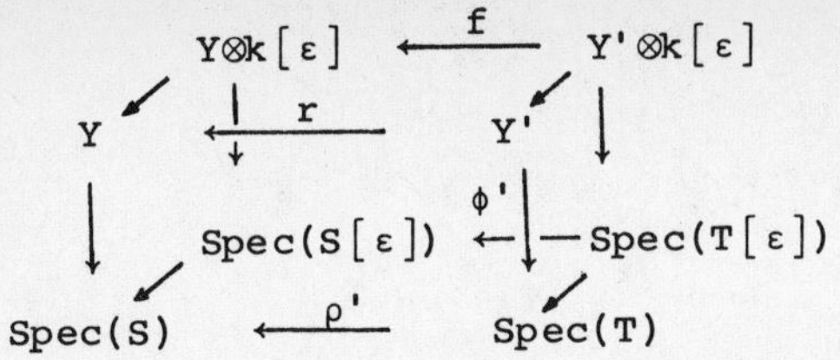

$$\begin{array}{ccccccc}
 & & Y\otimes k[\varepsilon] & \xleftarrow{\ f\ } & Y'\otimes k[\varepsilon] & & \\
 & \swarrow & \downarrow & & \swarrow & \downarrow & \\
Y & & \xleftarrow{\ r\ } & & Y' & & \\
\downarrow & & \mathrm{Spec}(S[\varepsilon]) & \xleftarrow{\ \phi'\ } & \downarrow & \mathrm{Spec}(T[\varepsilon]) & \\
 & \swarrow & & & \swarrow & & \\
\mathrm{Spec}(S) & & \xleftarrow{\ \rho'\ } & & \mathrm{Spec}(T) & &
\end{array}$$

Consider the set

$$V(\eta,\rho) = \{\phi' : \mathrm{Spec}(T[\varepsilon] \to \mathrm{Spec}(S[\varepsilon] \mid \text{there exists an } f \text{ making the diagram above commutative}\}.$$

Now, the morphisms $\phi' = \mathrm{Spec}(\phi) : \mathrm{Spec}(T[\varepsilon]) \to \mathrm{Spec}(S[\varepsilon])$ lifting $\mathrm{Spec}(\rho) : \mathrm{Spec}(T) \to \mathrm{Spec}(S)$, correspond to derivations $D \in \mathrm{Der}_k(S,T)$, such that $\phi = \rho + \varepsilon \cdot D$. Thus,

$$V(\eta,\rho) = \{D \in \mathrm{Der}_k(S,T) \mid (Y\otimes k[\varepsilon]) \otimes_{\phi} T[\varepsilon] \overset{\chi}{\simeq} Y'\otimes k[\varepsilon]$$

$$\text{where } \phi = \rho + \varepsilon \cdot D \text{ and } \chi \otimes_{k[\varepsilon]} k = \mathrm{id}_{Y'}\}.$$

Put $V(\eta) = V(\eta, \mathrm{id}_S)$. The following result is not entirely trivial

<u>Proposition (3.3)</u>. In the situation above

(i) $V(\eta,\rho) = \ker\{\mathrm{Der}_k(S,T) \xrightarrow{g_{\eta'}} A^1(S,Y;O_Y \otimes_S T)\}$

(ii) $V(\eta,\rho)$ is a T-submodule of $\mathrm{Der}_k(S,T)$

(iii) $V(\eta)$ is a sub Lie algebra of $\mathrm{Der}_k(S)$.

(iv) The natural morphisms of S-modules

$\mathrm{Der}_k(S,S) \to \mathrm{Der}_k(S,T) \xleftarrow{j} \mathrm{Der}_k(T,T)$

induce morphisms of S-modules

$$V(\eta) \to V(\eta,\rho) \leftarrow V(\eta')$$

Moreover $j^{-1}(V(\eta,\rho)) = V(\eta')$.

<u>Proof</u>. Let $d : S \to \Omega_{s/k}$ be the universal derivation, $d(s) = s\otimes 1 - 1\otimes s$, modulo I^2, and consider the morphisms

$$\nu_i : S \to S\otimes S/I^2 = S[\Omega_{s/k}],\ i = 1,2.$$

Clearly $\nu_1 - \nu_2 : S \to \Omega_{S/k}$ is equal to d. Let $D \in \mathrm{Der}_k(S,T)$ and consider D as an S-linear homomorphism $D' : \Omega_{S/k} \to T$. There is an associated map of Nagata rings

$$\rho - \varepsilon \cdot D' : S[\Omega_{S/k}] \to T[\varepsilon]$$

such that $(\rho-\varepsilon D')\circ\nu_1 = i\circ\rho : S \to T[\varepsilon]$ and $(\rho-\varepsilon D')\circ\nu_2 = \rho+\varepsilon\cdot D : S \to T[\varepsilon]$, where $i : T \to T[\varepsilon]$ is the obvious inclusion.

By definition of the Kodaira-Spencer map $g_{\eta'}$, we have

$$g_{\eta'}(D) = \partial_D(\nu_1^*(Y)-\nu_2^*(Y)) = (\rho-\varepsilon D')^*\nu_1^*(Y)-(\rho-\varepsilon D')^*\nu_2^*(Y)$$
$$= (i\circ\rho)^*(Y)-(\rho+\varepsilon D)^*(Y).$$

From this we see immediately that (i), and (iv) follow. Since $g_{\eta'}$ is T-linear (ii) is trivial. (iii) follows from the definition of $V(\eta)$, as $V(\eta)$ obviously is the Lie algebra of a corresponding group-functor, like $\ell(\pi)$ in §2. Q.E.D.

Lemma (3.4). Let S be a k-algebra (essentially) of finite type and let $\eta : Y \to \mathrm{Spec}(S)$ be a flat morphism of finite type. Denote by $\underline{m}$ a maximal ideal of S such that $S/\underline{m} = k$. Then for any coherent O_Y-module M,

(i) There is an exact sequence $0 \to \varprojlim_n{}^{(1)} A^{i-1}(S/\underline{m}^n, Y\otimes_S S/\underline{m}^n; M\otimes_S S/\underline{m}^n) \to$ $\to A^i(S^\wedge, Y^\wedge, M^\wedge) \to \varprojlim A^i(S/\underline{m}^n, Y\otimes_S S/\underline{m}^n; M\otimes_S S/\underline{m}^n) \to 0.$

(ii) $A^i(S^\wedge, Y^\wedge, M^\wedge) = A^i(S,Y,M)^\wedge$, $i\geq 0$.

(iii) If S is regular and M is flat as an S-module and R is an S-algebra of finite type, then there exists a spectral sequence given by:

$$E_2^{p,q} = \mathrm{Tor}^S_{-p}(A^q(S,Y;M),R)$$

converging to

$$A^{p+q}(S,Y;M\underset{S}{\otimes}R).$$

Proof. Suppose that $Y = \mathrm{Spec}(A)$. Recall the definition of the André cohomology,

$$A^q(S,A;M) = H^q(C^\cdot(M))$$

where $C^\cdot(M)$ is a functorial complex of A-modules such that $C^m(M) = \Pi_{I_m} M$ for some indexing set I_m.

It follows that

$$\varprojlim_n C^\cdot(M\otimes_S S/\underline{m}^n) = C^\cdot(M^\wedge)$$

$$\varprojlim_n{}^{(1)} C^\cdot(M\otimes_S S/\underline{m}^n) = 0$$

since the projective system $C^\cdot(M\otimes_S S/\underline{m}^n)$ is surjective. (i) follows therefore from general nonsense on $\varprojlim$ of complexes, see e.g. [La1], and [An]. (ii) follows from (21.2) of [An]. Let $L.$ be a finite S-free resolution of R, then $C^\cdot(M)\otimes_S L. = C^\cdot(M\otimes_S L.)$, and (iii) is simply the spectral sequence associated to the double complex $C^\cdot(M)\otimes_S L.$. The generalization to the global case presents no difficulties, see [La1], Chap 3. Q.E.D.

Suppose from now on that,

(A_1) there exists an algebraization $\pi:\tilde{X} \to \underline{H} = \mathrm{Spec}(H)$ of the formal versal family $\hat{\pi}:\hat{X}\to\hat{\underline{H}}$.

Consider the Kodaira-Spencer map $g_X = g_\pi$,

$$g_X\colon \mathrm{Der}_k(H) \to A^1(H,\tilde{X};O_{\tilde{X}})$$

and put

$$V = V(\pi) = \ker g_X.$$

Let $\rho' = \mathrm{Spec}(\rho):\underline{S} \to \underline{H}$ be a morphism, $\underline{s}_0 \in \underline{S}$ a closed point mapping to the base point $* \in \underline{H}$.

Denote by $\pi':\tilde{X}' \to \underline{S}$ the pull-back of π to $\underline{S}$, and consider $V(\pi,\rho) \subseteq \mathrm{Der}_k(H,S)$. Notice that $V(\pi,\rho) \subseteq \mathrm{Der}_k^c(H,S)$. This follows from the fact that $\hat{H}_*$ prorepresents the deformation functor on the subcategory $\underline{\ell}_2$ of $\underline{\ell}$. More presicely, if $D \in V(\pi,\rho)$ and if we let ϕ be the composition of $H \to H[\varepsilon]$ and $\rho+\varepsilon D:H[\varepsilon] \to S[\varepsilon]$, then $\tilde{X}\otimes_\phi S[\varepsilon] \simeq \tilde{X}\otimes_H S[\varepsilon]$. Let $\bar{\phi}$ be the composition of ϕ with $S[\varepsilon] \to k[\varepsilon]$ given by the point $\underline{s}_0$, then $\tilde{X}\otimes_{\bar{\phi}} k[\varepsilon] \simeq \tilde{X}\otimes_H k[\varepsilon]$ implying that $\bar{\phi}$ is equal to the composition $H \to k \to k[\varepsilon]$ defined by $*$, thus that D maps the maximal ideal $\underline{m}_*$ of H into the maximal ideal $\underline{n}_{\underline{s}_0}$ of S.

Corollary (3.5). With the assumptions and notations above we have

(i) $V(\pi,\rho)^{\wedge} = V(\pi^{\wedge},\rho^{\wedge}) = \ell(\pi^{\wedge},\rho^{\wedge})$.
where completion is with respect to the $\underline{n}_{\underline{s}_0}$ - resp. $\underline{m}_*$-adic topology.

(ii) $V(\pi,\rho)\otimes_H k \simeq A^0(k,X;O_X)/A^0_{\pi'}$.

(iii) $V(\pi)\otimes_H S \to V(\pi,\rho)$ is locally surjective.

Proof. Since we know that $V(\pi,\rho) \subseteq \mathrm{Der}^c_k(H,S)$ it is clear by (3.3) (i) that $V(\pi,\rho) = \ker\{\mathrm{Der}^c_k(H,S) \to A^1(S,\tilde{X}';O_{\tilde{X}'})\}$. Since by assumption the S-moduls involved are of finite type, completion with respect to the $\underline{n}$-adic topology commutes with the formation of kernel. Moreover $\mathrm{Der}^c_k(H,S)^{\wedge} \simeq \mathrm{Der}^c_k(H^{\wedge},S^{\wedge})$ and $A^1(S,\tilde{X}';O_{\tilde{X}'})^{\wedge} \simeq A^1(S^{\wedge},\tilde{X}'^{\wedge};O_{\tilde{X}'^{\wedge}})$ by (3.4), (ii). But this implies that $V(\pi,\rho)^{\wedge} = V(\pi^{\wedge},\rho^{\wedge})$ which is equal to $\ell(\pi^{\wedge},\rho^{\wedge})$. From this (i) follows. (ii) is a consequence of (i) and (2.6) (iv), and (iii) follows from (ii) and Nakayamas lemma.

Q.E.D.

Definition (3.6). A flat family $\pi:\tilde{X} \to \underline{S}$ is called formally versal if at every k-point $\underline{s}\in\underline{S}$, $g_{\underline{s}}:t_{\underline{S},\underline{s}} \to A^1(k(\underline{s}),X(\underline{s});O_{X(\underline{s})})$ is surjective.

Corollary (3.7). If a deformation of X,

$$\begin{array}{ccc} X & \to & \tilde{X} \\ \downarrow & & \downarrow \pi \\ * = \mathrm{Spec}(k) & \to & \underline{S} \end{array}$$

satisfies the conditions

a) $\underline{S}$ is smooth

b) $g_*: t_{\underline{S},*} \to A^1(k,X;O_X)$ is surjective

c) $A^i(S,\tilde{X};O_{\tilde{X}})$ is an $O_{\underline{S}}$-module of finite type for $i = 1$ and flat for $i \geq 2$,

then in some neighbourhood of $*$, π is formally versal

Proof. Consider the Kodaira-Spencer map g and the diagram

$$\begin{array}{ccc} \mathrm{Der}_k(S) & \xrightarrow{g} & A^1(S,\tilde{X};O_{\tilde{X}}) \\ \eta \downarrow & & \downarrow \mu \\ t_{\underline{S},*} & \xrightarrow{g_*} & A^1(k,X;O_X) \end{array}$$

By (3.4) (iii), and c), $A^1(S,\tilde{X};O_{\tilde{X}})\otimes_S k(\underline{s}) \simeq A^1(k(\underline{s}),X(\underline{s}),O_{X(\underline{s})})$ for all $\underline{s} \in \underline{S}$. Since moreover, by assumption, $A^1(S,\tilde{X};O_{\tilde{X}})$ is of finite type as S-module, it follows from Nakayamas lemma, that im g generates $A^1(S,\tilde{X};O_{\tilde{X}})$ in some neighbourhood of $*$. Then, by (3.2) $g_{\underline{s}}$ is surjective, for all $\underline{s}$ in some neighbourhood of $*$.

Q.E.D.

Suppose from now on that

(V) π is formally versal.

In particular it follows from (3.6) that this last assumption holds in some neighbourhood of $*$ if:

(V') (1) H is k-smooth
(2) $A^1(H,\tilde{X};O_{\tilde{X}})$ is an H-module of finite type.
(3) $A^i(H,\tilde{X};O_{\tilde{X}})$ is a flat H-module for $i \geq 2$.

Now, let $\underline{s} \in \mathrm{Spec}(S)$ be any k-rational point, then by (3.2) and the condition (V), there is a commutative diagram where the top and the bottom sequences are exact,

$$\begin{array}{ccccccc}
0 \to V(\pi,\rho) & \to & \mathrm{Der}_k(H,S) & \xrightarrow{g} & A^1(H,\tilde{X};O_{\tilde{X}}\otimes_H S) & & \\
\downarrow & & \downarrow & & \downarrow & & \\
0 \to V(\pi,\rho)\otimes_S k(\underline{s}) & \to & \mathrm{Der}_k(H,S)\otimes_S k(\underline{s}) & \to & A^1(H,\tilde{X},O_{\tilde{X}}\otimes_H S)\otimes_S k(\underline{s}) & \to & 0 \\
 & & m_{\underline{s}} \downarrow & & \downarrow \ell_{\underline{s}} & & \\
 & & t_{\underline{H},\rho(\underline{s})} & \to & A^1(k(\underline{s}),X(\rho(\underline{s}));O_{X(\rho(\underline{s}))}) & \to & 0
\end{array}$$

Remark (3.8). Observe that from this follows

$$\dim_k A^1(k,X(\underline{t});O_{X(\underline{t})}) = \dim t_{\underline{H},\underline{t}} - \mathrm{rank}_{\underline{S}} V(\pi,\rho)$$

for $\underline{t} = \rho'(\underline{s})$, whenever $m_{\underline{s}}$ and $\ell_{\underline{s}}$ are isomorphisms. Notice that $m_{\underline{s}}$ is an isomorphism when $\underline{t}$ is a non-singular point. In particular, we find in this case,

$$\min_{\underline{s} \in \underline{S}} \{\dim_k A^1(k,X(\rho'(\underline{s})),O_{X(\rho'(\underline{s}))})\} \geq \dim \underline{H} - \mathrm{rank}\ V(\pi,\rho).$$

Now, in the light of (3.5) (ii) above, we would like to know when $m_{\underline{s}}$ and $\ell_{\underline{s}}$ are isomorphisms and when $V(\pi,\rho)$ is locally free at $\underline{s}_0$. The following proposition is a partial result in this direction.

<u>Proposition (3.9)</u>. Assume either

(1) $\underline{S}$ is a non-singular curve.

or

(2) $\underline{S}$ is non-singular, and the conditions (V') hold.

Then,

(i) $\ell_{\underline{s}}$ and $m_{\underline{s}}$ are injective

and

(ii) in a neighbourhood of $\underline{s}_0$, $V(\pi,\rho)$ is locally free.

<u>Proof</u>. We have, by definition an exact sequence of S-modules

$$0 \to V(\pi,\rho) \to \mathrm{Der}_k(H,S) \to A^1(H_1,\tilde{X};O_{\tilde{X}_H}\otimes S).$$

Suppose first that $\underline{S}$ is a non-singular curve. S is a Dedekind domain. Since $\mathrm{Der}_k(H,S)$ has no S-torsion, $V(\pi,\rho)$ has no S-torsion. But then $\mathrm{Der}_k(H,S)$ and $V(\pi,\rho)$ are S-flat, and therefore locally free in a neighbourhood of $\underline{s}_0$.

Let $\underline{n}_{\underline{s}} = (v)$ be the maximal ideal of $\underline{s}\in\underline{S}$, and consider the exact sequence $0\to S \overset{v}{\to} S\to k(\underline{s})\to 0$. The induced sequence

$$\cdots \to A^i(H,\tilde{X};O_{\tilde{X}}\otimes_H S) \overset{v}{\to} A^i(H,\tilde{X};O_{\tilde{X}}\otimes_H S) \to A^i(k(\underline{s}),X(\underline{s});O_{X(\underline{s})}) \to \cdots$$

is therefore exact for all $i\geq 0$ and this implies that $m_{\underline{s}}$ and $\ell_{\underline{s}}$ are injective.

In case (2), observe that g is surjective in a neighbourhood of $*$, so that for any i, $\mathrm{Tor}_i^S(V(\pi,\rho),k)$ is a quotient of $\mathrm{Tor}_{i+1}^S(A^1(H,\tilde{X};O_{\tilde{X}}\otimes_H S),k)$. It suffices for our purpose to prove that $\mathrm{Tor}_2^S(A^1(H,\tilde{X};O_{\tilde{X}}\otimes_H S),k) = 0$. Now since $\pi':\tilde{X}'\to\underline{S}$ is the pull-back of π, we have $A^1(H,\tilde{X};O_{\tilde{X}}\otimes_H S) = A^1(S,\tilde{X}';O_{\tilde{X}'})$. Use the spectral sequence of (3.4) (iii). See first that the map

$$A^0(S,\tilde{X}';O_{\tilde{X}'})\otimes_S k = A^0_{\pi'} \overset{i}{\to} A^0(k,X;O_X)$$

is injective. This implies that the second differential in the spectral sequence,

$$d_2:\mathrm{Tor}_2^S(A^1(S,\tilde{X}';O_{\tilde{X}'}),k) \to A^0(S,\tilde{X}';O_{\tilde{X}'})\otimes_S k$$

is zero. For degree reasons all other differentials

$$d_k : \mathrm{Tor}_2^S(A^1(S,\tilde{X}';O_{\tilde{X}'}),k) \to \mathrm{Tor}_{2-k}^S(A^{1-k+1}(S,\tilde{X}';O_{\tilde{X}'}),k)$$

must vanish.

Now, since the abutment of the spectral sequence of total degree -1 is $A^{-1}(k,X;O_X) = 0$, so $E_\infty^{-2,1} = 0$ we find that $\mathrm{Tor}_2^S(A^1(S,\tilde{X}';O_{\tilde{X}'}),k) = E_2^{-2,1}$ must be the sum of the images

$$d_r : E_r^{-r-2,r} \to E_r^{-2,1}, \quad r \geq 2.$$

However, since by the assumption (V'), $\mathrm{Tor}_p^S(A^q(S,\tilde{X}';O_{\tilde{X}'}),-) = 0$ for all $p \geq 1$, $q \geq 2$ we find that $E_2^{-r-2,r} = 0$ for $r \geq 2$ therefore also $E_r^{-r-2,r} = 0$ for $r \geq 2$. Consequently $\mathrm{Tor}_2^S(A^1(S,\tilde{X}';O_{\tilde{X}'}),k) = 0$ and thus $\mathrm{Tor}_1(V(\pi,\rho),k) = 0$, i.e. $V(\pi,\rho)$ is locally free in a neighbourhood of s_0. Since $\underline{t} = \rho'(\underline{s})$ is a non-singular point, $\underline{m}_{\underline{s}}$ is an isomorphism and therefore also $\ell_{\underline{s}}$.

Q.E.D.

From this follows,

<u>Corollary (3.10)</u> (Wahls conjecture). Suppose there exists a non-singular curve $\underline{S}$ and a morphism $\rho' : \underline{S} \to \underline{H}$ such that the image of ρ' contains a non-singular point of $\underline{H}_{red}$ and such that the pull-back π' of π contains a rigid fiber. Then the dimension of the rigidifying component of $\underline{H}$ containing $\mathrm{im}\,\rho'$ is

$$\dim_k(A^0(k,X;O_X)/A^0_{\pi'})$$

<u>Proof</u>. This follows immediately from (3.9) (1) together with (3.5) and (3.8).

Q.E.D.

This result was conjectured by J. Wahl and proved by him in some special cases, see [W], and has recently also been proved in the complex analytic case by Greuel & Looijenga [G-L.].

Since by (2.9) we know how to compute the linearized action of V on $\underline{H}$, and since by [La1] all components of $\underline{H}$ have dimension greater or equal to

$$\dim_k A^1(k,X;O_X) - \dim_k A^2(k,X;O_X)$$

we obtain a smoothing, and a non-smoothing criterion as follows.

Proposition (3.11). Assume (A_1) and (V). Then we have:

(i) Suppose there exists an unobstructed $\xi \in A^1(k,X;O_X)$ such that $A^0(k,X;O_X)\cdot\xi = A^1(k,X;O_X)$, then X may be rigidifyed.

(ii) Let A^0_π be the sub Lie algebra of those elements of $A^0(k,X;O_X)$ that may be lifted everywhere. Suppose

$$\dim_k A^1(k,X;O_X) - \dim_k A^2(k,X;O_X) > \dim_k (A^0(k,X;O_X)/A^0_\pi)$$

then X cannot be rigidifyed.

Proof. (i) Let $\rho':\underline{S} \to \underline{H}$ be a morphism, $\underline{S}$ a non-singular curve, such that ξ is in the image of the tangent map $t_\rho : t_{\underline{S},\underline{s}_0} \to t_{\underline{H},*} \simeq A^1(k,X;O_X)$. By (2.8) we know that the orbit of ξ under $\ell(\pi)$ or, what is the same, under $A^0(k,X;O_X)$, is isomorphic to $A^0(k,X;O_X)/A^0_\xi$. Therefore $A^0(k,X;O_X)/A^0_\xi \simeq A^1(k,X;O_X)$. Now, obviously $A^0(k,X;O_X)/A^0_{\pi'} \to A^0(k,X;O_X)/A^0_\xi$ is surjective. By (3.9) $V(\pi,\rho)$ is locally free of rank $\dim_k(A^0(k,X;O_X)/A^0_{\pi'})$ at $\underline{s}_0$. Moreover, since $\underline{S}$ is a non-singular curve, there exist an open subset of points $\underline{s} \in \underline{S}$, such that the composition

$$V(\pi,\rho)\otimes_S k(\underline{s}) \to \mathrm{Der}_k(H,S)\otimes_S k(s) \xrightarrow{m_{\underline{s}}} t_{H,\rho'(\underline{s})}$$

is injective. Since $\dim t_{H,\rho'(\underline{s})} \leq \dim_k A^1(k,X;O_X)$ and since by the above, $\dim_k V(\pi,\rho)\otimes_S k(\underline{x}) \geq \dim_k A^1(k,X;O_X)$, we find $V(\pi,\rho)\otimes_S k(\underline{s}) = t_{\underline{H},\rho'(\underline{s})}$. But then the map $t_{H,\underline{t}} \to A^1(k(\underline{t}),X(\underline{t});O_{X(\underline{t})})$, where $\underline{t} = \rho'(\underline{s})$, is zero. However, by (V) this map is also surjective, therefore $X(\underline{t})$ is rigid.

(ii) is already proved. Q.E.D.

Proposition (3.12). (i) Suppose the conditions (V') hold. Let $\underline{H}_0$ be the subscheme of $\underline{H}$ defined by $\underline{V}(\pi) = 0$. Then for every point $\underline{t} \in \underline{H}_0$, the formalization $\hat{\underline{H}}_{0\underline{t}}$ of $\underline{H}_0$ at $\underline{t}$ is the prorepresenting substratum of the formal moduli $\hat{\underline{H}}(\underline{t})$ of $X(\underline{t})$.

(ii) $V(\pi)\otimes_{\underline{H}}\underline{H}_0$ is an $\underline{H}_0$-Lie algebra and a deformation of the Lie algebra $L(X)\simeq A^0(k,X;O_X)/A^0_\pi$.

<u>Proof</u>. The conditions (V') imply that for every $\underline{t}\in\underline{H}_0$, $\underline{H}^\wedge_{\underline{t}}\simeq H(\underline{t})^\wedge$. But then (3.12) follows from (3.5), with $\rho = \mathrm{id}_H$, together with (2.6) (v). Q.E.D.

Consider the H-module $A^1(H,\tilde{X};O_{\tilde{X}})$, and let $\{\underline{S}_\tau\}$, $\tau = 0,\dots,\tau^1 = \dim_k A^1(k,X;O_X)$ be the flattening stratification of $A^1(H,\tilde{X};O_{\tilde{X}})$, (see [M], Lecture 8), and let $\underline{S}_\tau = \bigcup_{c=1}^{k_\tau}\underline{S}_{\tau,c}$ be the decomposition of $\underline{S}_\tau$ into its connected components. We shall consider only those $\underline{S}_{\tau,c}$ for which $*\in\bar{\underline{S}}_{\tau,c}$.

Notice that the $\underline{S}_{\tau,c}$'s are locally closed subschemes of $\underline{H}$ and that $\underline{S}_{\tau^1} = \underline{H}_0$. Denote by π_τ, respectively $\pi_{\tau,c}$, the restriction of π to $\underline{S}_\tau$, respectively $\underline{S}_{\tau,c}$.

Put for every $\underline{t}\in\underline{H}$, with $\underline{t}\in\underline{S}_{\tau,c}$

$$T_{\underline{t}} = \bar{\underline{S}}_{\tau,c}\ .$$

Consider also, for any k-point $\underline{t}\in\underline{S}_\tau\subseteq\underline{H}$, the formal family

$$\pi^\wedge_{\underline{t}}:\ \tilde{X}^\wedge_{\underline{t}}\ \to\ \underline{H}^\wedge_{\underline{t}}$$

where $\underline{H}^\wedge_{\underline{t}} = \mathrm{Spf}(H^\wedge_{\underline{m}_{\underline{t}}})$, $\underline{m}_{\underline{t}}$ being the maximal ideal of H corresponding to $\underline{t}$.

As above we denote by $X(\underline{t})$ the fiber $\pi^{-1}(\underline{t})$. Obviously the closed fiber of $\pi^\wedge_{\underline{t}}$ is precisely $X(\underline{t})$.

Consider the formal moduli $\underline{H}(\underline{t})^\wedge$ of $X(\underline{t})$, i.e. $\underline{H}(\underline{t})^\wedge = \mathrm{Spf}(H(\underline{t})^\wedge)$, and let

$$\pi(\underline{t}):\ \tilde{X}(\underline{t})\ \to\ \underline{H}(\underline{t}),$$

with $\underline{H}(\underline{t}) = \mathrm{Spec}(H(\underline{t}))$, be an algebraization of the formal versal family

$$\pi(\underline{t})^\wedge:\ X(\underline{t})^\wedge\ \to\ \underline{H}(\underline{t})^\wedge.$$

Let $\underline{H}_0(\underline{t})$ be the prorepresenting substratum of $\underline{H}(\underline{t})$, and let

$$\pi_0(\underline{t}):\tilde{X}_0(\underline{t}) \to \underline{H}_0(\underline{t})$$

be the restriction of $\pi(\underline{t})$. Then, by formal versality, there is a morphism

$$\hat{\rho}_{\underline{t}}: \hat{\underline{H}}_{\underline{t}} \to \underline{H}(\underline{t})^{\wedge}$$

such that $\hat{\pi}_{\underline{t}}$ is the pull-back of $\pi(\underline{t})^{\wedge}$ by $\hat{\rho}_{\underline{t}}$. Since π is versal, $\hat{\rho}_{\underline{t}}$ has maximal rank at $\underline{t}$.

Now to proceed we have to assume that our objects X satisfy the condition:

(A_2) If $\eta:Y \to \underline{S}$ is a flat family with fiber $X = \eta^{-1}(\underline{s})$ for some closed point $\underline{s} \in \underline{S}$, then there exists an étale neighbourhood $\varepsilon:\underline{E} \to \underline{S}$ of $\underline{s}$ and a morphism $\rho:\underline{E} \to \underline{H}$ such that there is an isomorphism between the pull-back $\varepsilon^*(\eta)$ of η and the pull-back $\rho^*(\pi)$ of π.

In the applications we have in mind (A_2) is a consequence of M. Artins approximation theorem, see [Ar].

Now, (A_2) implies the existence of an étale neighbourhood

$$\eta(\underline{t}): \underline{E}(\underline{t}) \to \underline{H}$$

of $\underline{t} \in \underline{H}$, and a dominant morphism

$$\rho(\underline{t}): \underline{E}(\underline{t}) \to \underline{H}(\underline{t})$$

such that $\eta(\underline{t})^*(\pi) = \rho(\underline{t})^*(\pi(\underline{t}))$.

Put $\underline{E}_\tau(\underline{t}) = \rho(\underline{t})^{-1}(\underline{H}_0(\underline{t}))$ and consider the cartesian diagram

$$\begin{array}{ccc} \underline{E}(\underline{t}) & \xrightarrow{\rho(\underline{t})} & \underline{H}(\underline{t}) \\ \uparrow & & \uparrow \\ \underline{E}_\tau(\underline{t}) & \xrightarrow{\rho_\tau(\underline{t})} & \underline{H}_0(\underline{t}) \end{array}$$

and let

$$\pi'(\underline{t}):\tilde{Y}(\underline{t}) \to \underline{E}(\underline{t})$$

$$\pi'_\tau(\underline{t}):\tilde{Y}_\tau(\underline{t}) \to \underline{E}_\tau(\underline{t})$$

be the corresponding pull-backs of the family π.

It is clear that $\rho(\underline{t})$ is of constant maximal rank at every closed point of $\underline{E}_\tau(\underline{t})$. Assume for a moment that for every $\underline{t} \in \underline{H}$, $\underline{H}(t)$ is non-singular and in particular that $\underline{H}$ is non-singular, then $E(\underline{t})$ is non-singular, and $\rho(\underline{t})$ is smooth in a neighbourhood of every point of $\underline{E}_\tau(\underline{t})$.

Consequently $\rho_\tau(\underline{t})$ is smooth. In particular the fibers of $\rho_\tau(\underline{t})$ are smooth subschemes of $\underline{E}_\tau(\underline{t})$. It is clear that any such fiber is a maximal integral submanifold for $V(\pi'_\tau)$, see remark (3.8). Let us denote by $\underline{D}(\underline{t})$ the special fiber of $\rho_\tau(\underline{t})$.

<u>Corollary (3.13)</u>. Suppose the conditions (V') hold. Through every point $\underline{t} \in \underline{H}$ there passes a smooth maximal integral submanifold $\underline{D}_{\underline{t}}$ for V. Moreover $\underline{D}_{\underline{t}} \subset \underline{T}_{\underline{t}}$ and

$$\dim \underline{H}_0(\underline{t}) = \dim \underline{T}_{\underline{t}} - \dim \underline{D}_{\underline{t}}.$$

<u>Proof</u>. Recall that (V') implies that $\underline{H}(\underline{t})$ is non-singular for every $\underline{t} \in \underline{H}$. Let $\underline{D}_{\underline{t}} = \eta(\underline{t})(\underline{D}(\underline{t}))$ and glue. Q.E.D.

Let $\underline{H}_0(\underline{t}) = \mathrm{Spec}(H_0(\underline{t}))$, $\underline{E}(\underline{t}) = \mathrm{Spec}(E(\underline{t}))$, $\underline{E}_\tau(t) = \mathrm{Spec}(E_\tau(\underline{t}))$. By shrinking the étale neighbourhood $\underline{E}(\underline{t})$ of $\underline{t} \in \underline{H}$ we may assume that $\rho(\underline{t})$ is smooth.

<u>Lemma (3.14)</u>. In the above situation, there exists an exact sequence of $E(\underline{t})$-modules

$$0 \to \mathrm{Der}_{H(\underline{t})}(E(\underline{t}),E(\underline{t})) \to V(\pi'(\underline{t})) \to V(\pi(\underline{t})) \underset{H(\underline{t})}{\otimes} E(\underline{t}) \to 0.$$

<u>Proof</u>. Since $E(\underline{t})$ is $H(\underline{t})$-smooth, the sequence

$$0 \to \mathrm{Der}_{H(\underline{t})}(E(\underline{t}),E(\underline{t})) \xrightarrow{i} \mathrm{Der}_k(E(\underline{t}),E(\underline{t})) \xrightarrow{j} \mathrm{Der}_k(H(\underline{t}),E(\underline{t})) \to 0$$

is exact. Now, use (3.3) and (3.5) for the situation

$$\begin{array}{ccc} \tilde{X}(\underline{t}) & \leftarrow & \rho(\underline{t})^*(\tilde{X}(\underline{t})) = \tilde{Y}(\underline{t}) \\ \downarrow\pi(\underline{t}) & & \downarrow\pi'(\underline{t}) \\ \underline{H}(\underline{t}) & \underset{\rho(\underline{t})}{\leftarrow} & \underline{E}(\underline{t}) \end{array}$$

By (3.3) (iv) we know that $V(\pi'(\underline{t})) = j^{-1}(V(\pi(\underline{t}),\rho))$ and from (3.5) (iii) follows that $V(\pi(\underline{t})) \otimes_{H(\underline{t})} E(\underline{t}) \to V(\pi(\underline{t}),\rho)$ is surjective. However, since $E(\underline{t})$ is $H(\underline{t})$-flat and $\mathrm{Der}_k(H(\underline{t})) \otimes_{H(\underline{t})} E(\underline{t}) \simeq \mathrm{Der}_k(H(\underline{t}),E(\underline{t}))$, $V(\pi(\underline{t})) \otimes_{H(\underline{t})} E(\underline{t}) \to V(\pi(t),\rho)$ is also injective, therefore an isomorphism. Q.E.D.

Notice that since $\underline{E}_\tau(\underline{t}) \to \underline{H}_0(\underline{t})$ is smooth and $V(\pi_0(\underline{t})) = 0$ the above proof yields the formula

$$V(\pi'_\tau(\underline{t})) = \mathrm{Der}_{H_0(\underline{t})}(E_\tau(\underline{t})).$$

Observe also that $E_\tau(\underline{t})$ is stable under $V_{E(\underline{t})}$. To see this let $\delta \in V_{E(\underline{t})}$, and consider the automorphism

$$\kappa = \mathrm{id}+\varepsilon\delta : E(\underline{t})\otimes k[\varepsilon] \to E(\underline{t})\otimes k[\varepsilon].$$

Since the corresponding automorphism $\kappa_*^{-1}\kappa^*$ of $\mathrm{Der}_k(E(\underline{t}))\otimes k[\varepsilon]$ is given by $\kappa_*^{-1}\kappa^*(D) = D+\varepsilon[D,\delta]$, it maps $V(\pi'(\underline{t})\otimes k[\varepsilon])$ into itself. Consequently in $E(\underline{t})\otimes k[\varepsilon]$ we must have that $\mathrm{rank}\{\delta_1,\ldots,\delta_{r+1}\} = \mathrm{rank}\{\delta_1+\varepsilon[\delta_1,\delta],\ldots,\delta_{r+1}+\varepsilon[\delta_{r+1},\delta]\}$ for all sequences $\delta_1,\ldots,\delta_{r+1}$ from $V(\pi'(\underline{t}))$. Let $\mathfrak{a}_\tau$ be the ideal generated by the determinants of the form

$$\det\begin{pmatrix} \delta_1(a_1), & \delta_1(a_2), & \ldots, & \delta_1(a_{r+1}) \\ \cdot & \cdot & \cdot & \cdot \\ \delta_{r+1}(a_1), & \delta_{r+1}(a_2), & \ldots, & \delta_{r+1}(a_{r+1}) \end{pmatrix}, \quad \delta_i \in V_{E(\underline{t})},\ a_j \in E(\underline{t}),\ r=\tau^1-\tau$$

Since $\mathfrak{a}_\tau$ defines the closure of the flattening stratum $E_\tau(\underline{t})$ of $A^1(E(\underline{t}),\tilde{Y}(\underline{t}),O_{\tilde{Y}(\underline{t})})$, the ideal in $E(\underline{t})\otimes k[\varepsilon]$ generated by $\mathfrak{a}_\tau$ must be equal to the ideal generated by the determinants of the form

$$\det\begin{pmatrix} \delta_1(a_1)+\varepsilon[\delta_1,\delta](a_1), & \ldots, & \delta_1(a_{r+1})+\varepsilon[\delta_1,\delta](a_{r+1}) \\ \vdots & & \\ \delta_{r+1}(a_1)+\varepsilon[\delta_{r+1},\delta](a_1), & \ldots, & \delta_{r+1}(a_{r+1})+\varepsilon[\delta_{r+1},\delta](a_{r+1}) \end{pmatrix}$$

$$= \varepsilon\sum_j \det\begin{pmatrix} \delta_1(a_1),\ldots,[\delta_1,\delta](a_j),\ldots,\delta_1(a_{r+1}) \\ \\ \delta_{r+1}(a_1),\ldots,[\delta_{r+1},\delta](a_j),\ldots,\delta_{r+1}(a_{r+1}) \end{pmatrix}$$

$$+ \det\begin{pmatrix} \delta_1(a_1),\ldots,\delta_1(a_{r+1}) \\ \\ \delta_{r+1}(a_1),\ldots,\delta_{r+1}(a_{r+1}) \end{pmatrix}$$

Consequently

$$\delta \det \begin{pmatrix} \delta_1(a_1), & \delta_1(a_2), \ldots, & \delta_1(a_{r+1}) \\ \vdots & & \\ \delta_{r+1}(a_1), & \delta_{r+1}(a_2), \ldots, & \delta_{r+1}(a_{r+1}) \end{pmatrix}$$

$$= \sum_j \det \begin{pmatrix} \delta_1(a_1), \ldots, & [\delta,\delta_1](a_j), \ldots, & \delta_1(a_{r+1}) \\ \vdots & \vdots & \\ \delta_{r+1}(a_1), \ldots, & [\delta,\delta_{r+1}](a_j), \ldots, & \delta_{r+1}(a_{r+1}) \end{pmatrix}$$

$$+ \sum_j \det \begin{pmatrix} \delta_1(a_1), \ldots, & \delta_1(\delta(a_j)), \ldots, & \delta_1(a_{r+1}) \\ \vdots & \vdots & \vdots \\ \delta_{r+1}(a_1), \ldots, & \delta_{r+1}(\delta(a_j)), \ldots, & \delta_{r+1}(a_{r+1}) \end{pmatrix}$$

is contained in the ideal $\mathfrak{a}_\tau$. Thus δ maps $\mathfrak{a}_\tau$ into itself, i.e. $E_\tau(\underline{t})$ is stable under $V(\pi'(\underline{t}))$. In exactly the same way we prove that $\underline{S}_\tau$ is stable under $V(\pi)$.

From this follows that the image of $V(\pi)$ (resp. $V(\pi'(\underline{t}))$) by the map $\mathrm{Der}_k(H,H) \to \mathrm{Der}_k(H,O_{\underline{S}_\tau})$ (resp. $\mathrm{Der}_k(E(\underline{t}),E(\underline{t})) \to \mathrm{Der}_k(E(\underline{t}),E_\tau(\underline{t}))$) sits in $\mathrm{Der}(O_{\underline{S}_\tau},O_{\underline{S}_\tau}) \subseteq \mathrm{Der}(H,O_{\underline{S}_\tau})$ (resp. $\mathrm{Der}_k(E_\tau(\underline{t}),E_\tau(\underline{t})) \subseteq \mathrm{Der}_k(E(\underline{t}),E_\tau(\underline{t}))$). Moreover $V(\pi)\otimes_H O_{\underline{S}_\tau}$ and $V(\pi'(t)) \otimes_{E(\underline{t})} E_\tau(\underline{t})$ are k-Lie algebras.

<u>Lemma (3.15)</u>. Suppose the conditions (V') hold. Then there are surjective homomorphisms of Lie algebras

$$V(\pi'(\underline{t})) \to V(\pi'_\tau(\underline{t})), \quad V(\pi) \to V(\pi_\tau)$$

inducing surjective $E_\tau(\underline{t})$, resp. $O_{\underline{S}_\tau}$-module homomorphism

$$V(\pi'(\underline{t})) \otimes_{E(\underline{t})} E_\tau(\underline{t}) \to V(\pi'_\tau(\underline{t})), \quad V(\pi)\otimes_H O_{\underline{S}_\tau} \to V(\pi_\tau).$$

<u>Proof</u>. The existence is already clear. To prove the surjectivity it suffices, using (3.3) (iv) to show that $V(\pi'(\underline{t}))$ maps surjectively onto $V(\pi'(t),p)$ where $p:E(\underline{t}) \to E_\tau(\underline{t})$ is the quotient map. But

consider the diagram of exact sequences,

$$\begin{array}{ccccccc}
V(\pi'(\underline{t})) \underset{E(\underline{t})}{\otimes} E_\tau(\underline{t}) & \to & Der_k(E(\underline{t})) \underset{E(\underline{t})}{\otimes} E_\tau(\underline{t}) & \to & A^1(E(\underline{t}),\tilde{Y},O_{\tilde{Y}}) \underset{E(\underline{t})}{\otimes} E_\tau(\underline{t}) & \to & 0 \\
\downarrow s & & \downarrow r & & \downarrow q & & \\
0 \to V(\pi'(\underline{t}),p) & \to & Der_k(E(\underline{t}),E_\tau(\underline{t})) & \to & A^1(E_\tau(\underline{t}),\tilde{Y}_0,O_{\tilde{Y}_\tau}) & \to & 0
\end{array}$$

By (3.4) (iii) q is an isomorphism, r is surjective (in fact, r is an isomorphism) therefore s is surjective. Q.E.D.

Combining the last two lemmas, we obtain the following diagram of exact sequences,

$$\begin{array}{ccccccc}
 & & Der_{H(\underline{t})}(E(\underline{t}),E_\tau(\underline{t})) & & & & \\
 & & \downarrow & \searrow \ell(\underline{t}) & & & \\
0 \to \ker n(\underline{t}) & \to & V(\pi'(\underline{t})) \underset{E(\underline{t})}{\otimes} E_\tau(\underline{t}) & \xrightarrow{n(\underline{t})} & V(\pi'_\tau(\underline{t})) & \to & 0 \\
 & m(\underline{t}) \searrow & \downarrow & & & & \\
 & & V(\pi(\underline{t})) \underset{H(\underline{t})}{\otimes} E_\tau(\underline{t}) & & & & \\
 & & \downarrow & & & & \\
 & & 0 & & & &
\end{array}$$

from which we deduce

<u>Proposition (3.16)</u>. Suppose the conditions (V') hold. For every $\underline{t} \in \underline{S}_\tau$, $\ell(\underline{t})$ is an isomorphism, implying that

(i) $V(\pi(\underline{t})) \otimes_{H(\underline{t})} k(\underline{t}) \simeq \ker\{n(\underline{t}): V(\pi)\otimes_H k(\underline{t}) \to V(\pi_\tau)\otimes k(\underline{t})\}$

(ii) $V(\pi'(\underline{t})) \otimes_{E(\underline{t})} E_\tau(\underline{t})$ is a semidirect product of $(V(\pi(\underline{t})) \otimes_{H(\underline{t})} E_\tau(\underline{t}))$ and $Der_{H_0(\underline{t})}(E_\tau(\underline{t}))$, as Lie algebras

(iii) $V(\pi'_\tau(\underline{t})) \simeq Der_{H_0(\underline{t})}(E_\tau(\underline{t}),E_\tau(\underline{t}))$.

(iv) The kernel of $V(\pi)\otimes_H O_{\underline{S}_\tau} \to V(\pi_\tau)$ is a flat $O_{\underline{S}_\tau}$-Lie algebra, the fibers of which are the Lie algebras $L(\underline{t})$, $\underline{t} \in \underline{S}_\tau$.

Proof. Clearly $\mathrm{Der}_{H(\underline{t})}(E(\underline{t}),E_\tau(\underline{t})) \simeq \mathrm{Der}_{H_0(\underline{t})}(E_\tau(\underline{t}))$ which by the formula following (3.14) is equal to $V(\pi_\tau'(\underline{t}))$, therefore $\ell(\underline{t})$ is an isomorphism. Consequently $n(\underline{t})$ splits and (ii) follows. Tensorization with $k(\underline{t})$ on $T(\underline{t})$ yields (i). Finally (iii) is already proved, and (iv) follows from (ii) since $E_\tau(\underline{t})$ is faithfully flat as $O_{\underline{S}_\tau}$-module. Q.E.D

Remark (3.17). It follows from (3.16) that $V(\pi'(\underline{t})) \otimes_{E(\underline{t})} E_\tau(\underline{t})$ is the semidirect product of the $H_0(\underline{t})$-Lie algebras $(V(\pi(\underline{t})) \otimes_{H(\underline{t})} E_\tau(\underline{t})) = (V(\pi(\underline{t})) \otimes_{H(\underline{t})} H_0(\underline{t})) \otimes_{H_0(\underline{t})} E_\tau(\underline{t})$ and $\mathrm{Der}_{H_0(\underline{t})}(E_\tau(\underline{t}),E_\tau(\underline{t}))$. Consequently $V(\pi'(\underline{t})) \otimes_{E(\underline{t})} E_\tau(\underline{t})$ is an $H_0(\underline{t})$-Lie algebra, and $(V(\pi(\underline{t})) \otimes_{H(\underline{t})} E_\tau(\underline{t}))$ is an $E_\tau(\underline{t})$-Lie algebra.

Proposition (3.18). Assume the conditions (V') hold. Let $0 \leq \tau \leq \tau^1$. Then there is a unique way of glueing together the families

$$\pi_0(\underline{t}):\tilde{X}_0(\underline{t}) \to \underline{H}_0(\underline{t}),\ \underline{t} \in \underline{S}_\tau$$

in the category of algebraic spaces, to form a family

$$\bar{\pi}_\tau : \bar{X}_\tau \to \underline{M}_\tau$$

such that,

(i) $\underline{M}_\tau$ is a quotient of $\underline{S}_\tau$, i.e. such that there is a natural epimorphism $\underline{S}_\tau \to \underline{M}_\tau$.

(ii) There exists in the category of algebraic spaces a Cartesian square

$$\begin{array}{ccc} \tilde{X}_\tau & \to & \bar{X}_\tau \\ \pi_\tau \downarrow & \square & \downarrow \bar{\pi}_\tau \\ \underline{S}_\tau & \to & \underline{M}_\tau \end{array}$$

(iii) if $\bar{\pi}_\tau$ restricted to some subscheme $\underline{K} \subseteq \underline{M}_\tau$, is constant, then $\underline{K}$ is finite and reduced.

Proof. Take a finite covering of $\underline{S}_\tau$ in the étale topology, of the form $E_\tau(\underline{t}_i) \to \underline{S}_\tau$, and consider the diagram

$$\begin{array}{ccccc}
\coprod_{i,j} E_\tau(\underline{t}_i) \times_{S_\tau} E_\tau(\underline{t}_j) & \rightrightarrows & \coprod_e E_\tau(\underline{t}_e) & \to & \underline{S}_\tau \\
\downarrow \rho_1 & & \downarrow & & \\
\coprod_{i,j} \underline{H}_0(\underline{t}_i) \times \underline{H}_0(\underline{t}_j) & \rightrightarrows & \coprod_e \underline{H}_0(\underline{t}_e) & &
\end{array}$$

Consider the scheme theoretic image $\underline{R}$ of ρ_1. Obviously $\underline{R} \to \coprod_{ij} \underline{H}_0(\underline{t}_i) \times \underline{H}_0(\underline{t}_j) = U \times U$, where we have put $U = \coprod_i \underline{H}_0(\underline{t}_i)$, is a closed equivalence relation.

Now we shall prove that $\underline{R}$ is étale. This is clearly the same as to say that $\underline{R}_{ij} \subset \underline{H}_0(t_i) \times \underline{H}_0(\underline{t}_j)$ is étale on $\underline{H}_0(\underline{t}_k)$, $k = i,j$. Consider this on the affine level, then $R_{ij} = \mathrm{im}\{H_0(\underline{t}_i) \otimes H_0(\underline{t}_j) \to T(\underline{t}_i) \otimes_{S_\tau} T(\underline{t}_j)\}$, and we must show that for every k-point $\underline{t} \in \underline{R}_{ij}$ mapping to the point $\underline{t}_1 \in \underline{H}_0(\underline{t}_i)$ and to the point $\underline{t}_2 \in \underline{H}_0(\underline{t}_j)$ the homomorphisms

$$H_0(\underline{t}_i)^\wedge_{\underline{t}_1} \to R^\wedge_{ij,\underline{t}}$$

$$H_0(\underline{t}_j)^\wedge_{\underline{t}_2} \to R^\wedge_{ij,\underline{t}}$$

are isomorphisms.

Now there exist points $\underline{s}_1 \in E_\tau(\underline{t}_i)$, $\underline{s}_2 \in E_\tau(\underline{t}_j)$ mapping to the same point $\underline{s} \in \underline{S}$, such that $\underline{s}_e$ maps to $\underline{t}_e$, $e = 1,2$.
Consider the morphism of local rings

$$\begin{array}{ccccc}
 & E_\tau(\underline{t}_i)_{\underline{s}_1} & \leftarrow & H_0(\underline{t}_i)_{\underline{t}_1} & \\
\nearrow & & & & \nwarrow \phi_1 \\
S_{\underline{s}} & \leftarrow & \phi & & H_0(\underline{s})_* \\
\searrow & & & & \swarrow \phi_2 \\
 & E_\tau(\underline{t}_j)_{\underline{s}_2} & \leftarrow & H_0(\underline{t}_j)_{\underline{t}_2} &
\end{array}$$

with their induced families. Since the fiber of the family on $\underline{H}_0(\underline{t}_i)$ at $\underline{t}_1$ is isomorphic to the one on $\underline{H}_0(\underline{t}_j)$ at $\underline{t}_2$, and both are isomorphic to the special fiber of the family on $\underline{H}_0(\underline{s})$, we know that there must, after completion, exist isomorphisms ϕ_i, $i = 1,2$, and a morphism ϕ as in the diagram, compatible with the families. Since we are in the prorepresenting substratum all such morphisms must be unique. Consequently the following diagram must be commutative,

$$\begin{array}{ccccc}
 & E_\tau(\underline{t}_i)^\wedge_{\underline{s}_1} & \leftarrow & H_0(\underline{t}_1)^\wedge_{\underline{t}_i} & \\
\nearrow & & & & \nwarrow \overset{\phi_1}{\simeq} \\
S^\wedge_{\underline{s}} & & \xleftarrow{\quad\psi\quad} & & H_0(\underline{s})^\wedge_* \\
\searrow & & & & \swarrow \underset{\phi_2}{\simeq} \\
 & E_\tau(\underline{t}_j)^\wedge_{\underline{s}_2} & \leftarrow & H_0(\underline{t}_j)^\wedge_{\underline{t}_2} &
\end{array}$$

But then $H_0(\underline{t}_i)^\wedge_{\underline{t}_1} \otimes H_0(\underline{t}_j)^\wedge_{\underline{t}_2}$ maps onto $\psi(H_0(\underline{s})^\wedge_*)$ in $S^\wedge_{\underline{s}}$. Since $R^\wedge_{ij,\underline{t}}$ is equal to this image, we obtain isomorphisms

$$H_0(\underline{t}_i)^\wedge_{\underline{t}_1} \simeq R^\wedge_{ij,\underline{t}}$$

$$H_0(\underline{t}_j)^\wedge_{\underline{t}_2} \simeq R^\wedge_{ij,\underline{t}}$$

Since R_{ij}, and $H_0(\underline{t}_k)$, $k = i,j$, are of finite type as k-algebras and since k is separably closed the proposition follows from (EGA IV 17.6.3), see also [Kn] (I.4.5).

Q.E.D.

There are no reasons to expect the $\underline{M}_\tau$'s to be schemes, and there are, in particular, no reasons at all to expect $\underline{M}_\tau$ to be a geometric quotient of $\underline{S}_\tau$ by the action of V. In fact, we shall show by an example, see §7, that, in general, $\underline{S}_\tau/V_\tau$ does not exist as a geometric quotient. However we shall under the conditions (V'), prove that there is a scheme-theoretic quotient of an open dense subset of any reduced normalized component $\underline{S}_{\tau,c}$ of $\underline{S}_\tau$, having good properties. First we need some preparation.

<u>Lemma (3.19)</u>. Let S be a Noetherian normal domain, $K = K(S)$ its quotient field and let T be a subring of K containing S. Suppose for every prime ideal $\underline{p}$ of S of height 1 that $\mathrm{Spec}(T_{\underline{p}}) \to \mathrm{Spec}(S_{\underline{p}})$ is surjective. Then $S = T$.

<u>Proof</u>. Let $\underline{p}$ be any prime ideal of height 1 in S, and let $\underline{q}$ be any preimage in T. Then $S_{\underline{p}} \subseteq T_{\underline{q}} \subseteq K$. Since $S_{\underline{p}}$ is a valuation ring, therefore maximal, we have $S_{\underline{p}} = T_{\underline{q}}$. Obviously $T \subseteq \cap T_{\underline{q}} = \cap S_{\underline{p}} =$ $= S$, the last equality being well-known, see for instance, [Serre, J. P.: Multiplicités (III-13, Remarque à Proposition 9)]. Q.E.D.

Notice also the following classical result,

Lemma (3.20). If S is a normal domain with quotient field K, and if $K \subseteq L$ is a finite separable field extension, then the normal closure T of S in L is a finite extension of S.

Proof. See, Serre: Loc-cit. (III-16, Proposition 11) or Zariski-Samuel: Vol. 1, Ch. V, Thm. 9.

Q.E.D.

Lemma (3.21). Let $\pi_0':\tilde{X}_0' \to \underline{H}_0'$ be the restriction of π_0 to $\underline{H}_0' = H_0$. Suppose $\underline{H}_0'$ is reduced and irreducible and let $\underline{h}_0 \in \underline{H}_0$. Let $\rho_0:\underline{H}_1 \to \underline{H}_0$ be the normalization of $\underline{H}_0$, and let $\rho_1:\underline{N} \to \underline{H}_1$ be étale and finite. Consider an epimorphism $\rho:\underline{T} \to \underline{N}$ commuting with the automorphisms g on $\underline{N}$ and g' on $\underline{T}$. Let $\underline{t} \in \underline{T}$, $\underline{n} \in N$, $\underline{h} \in \underline{H}_1$ be such that $\rho(\underline{t}) = \underline{n}$, $\rho_1(\underline{n}) = \underline{h}$, $\rho_0(\underline{h}) = \underline{h}_0$. Suppose $\underline{n}$ is a fix-point for g, and assume there is an isomorphism $\theta_g':g'^*(\rho_0 \circ \rho_1 \circ \rho)^*(\tilde{X}_0) \tilde{\to} (\rho_0 \circ \rho_1 \circ \rho)^*(\tilde{X}_0)$. Then $g = id_{\underline{N}}$.

Proof. Since $\rho^* \rho_1^* \rho_0^*(\tilde{X}_0') \simeq g'^* \rho^* \rho_1^* \rho_0^*(\tilde{X}_0')$ at $\underline{t}$ it is clear that the formalization of $\rho_0 \circ \rho_1 \circ \rho$ at $\underline{t}$ resp. $\underline{h}_0$ coincides with the formalization of $\rho_0 \circ \rho_1 \circ \rho \circ g' = \rho_0 \circ \rho_1 \circ g \circ \rho$ at $\underline{t}$ resp. $\underline{h}_0$. Since ρ is epimorphic, the formalization of $\rho_0 \circ \rho_1$ at $\underline{n}$ resp. $\underline{h}_0$ coincides with that of $\rho_0 \circ \rho_1 \circ g$, but then ρ_1 and $\rho_1 \circ g$ coincides since ρ_0 is a normalization. But then $g \in \mathrm{Aut}_{H_1}(N)$ and since $\underline{n}$ is unramified over $\underline{h}$, g cannot leave $\underline{n}$ fixed unless g is the identity.

Q.E.D.

Lemma (3.22). Keep the notations and assumptions of (3.21). Assume moreover that $\rho:\underline{T} \to \underline{N}$ is flat. Then there is a unique isomorphism

$$\theta_g:g^*(\rho_0 \circ \rho_1)^*(\tilde{X}_0') \tilde{\to} (\rho_0 \circ \rho_1)^*(\tilde{X}_0')$$

commuting with θ_g'.

Proof. By assumption ρ is faithfully flat and the result therefore follows from ordinary descent theory, see [Gr2] (B. Theorem 2).

Q.E.D.

Remark (3.23). Instead of a family $\pi_0 : \tilde{X}_0 \to \underline{H}_0$ of preschemes, we may consider any family of quasicoherent sheaves. The conclusion of (3.21) and (3.22) still hold. From now on, we shall assume that the objects X we are considering is of a class for which faithfully flat morphisms are of descent and étale morphisms are of strict descent. It follows from [Gr2] (B. Theorem 3) that this includes the class of quasicoherent sheaves and the class of quasiprojective schemes.

Now we are ready to prove the main theorem of this §.

Theorem (3.24). Suppose the conditions (V') hold for all objects $X(\underline{t})$ involved. Suppose also that these objects satisfy the assumptions of (3.23). Let $\underline{S}^n_{\tau,c}$ be the normalization of the reduction of the component $\underline{S}_{\tau,c}$ of $\underline{S}_\tau$. Then there exists an open dense subscheme $\underline{S}'_{\tau,c}$ of $\underline{S}^n_{\tau,c}$ and a quotient

$\rho_{\tau,c} : \underline{S}'_{\tau,c} \to \underline{N}_{\tau,c}$ such that $\underline{N}_{\tau,c}$ is of finite type and such that $O_{\underline{N}_{\tau,c}}$ is the ring of invariants of $O_{\underline{S}'_{\tau,c}}$ under $V(\pi')$, where $\pi' : \tilde{X}'_{\tau,c} \to \underline{S}'_{\tau,c}$ is the pull-back of π_τ. In the category of algebraic spaces there is a Cartesian diagram

$$\begin{array}{ccc} \tilde{X}'_{\tau,c} & \xrightarrow{\pi'} & \underline{S}'_{\tau,c} \\ \downarrow & & \downarrow \rho_{\tau,c} \\ \mathcal{X}_{\tau,c} & \xrightarrow[\mu_{\tau,c}]{} & \underline{N}_{\tau,c} \end{array}$$

such that

(i) If $Y \to \underline{N}_{\tau,c}$ is a subscheme along which $\mu_{\tau,c}$ is constant, then Y is finite.

(ii) $\underline{N}_{\tau,c}$ is formally in the prorepresenting substratum, for all c, if and only if the reduction of every component of $H_0(\underline{s})$, $\underline{s} \in \underline{S}_\tau$, is normal.

(iii) $\rho_{\tau,c}$ is smooth.

(iv) There is a quasifinite dominant morphism $\underline{N}_{\tau,c} \to \underline{M}_\tau$.

Proof. Consider $\underline{S}^n_{\tau,c} \to \underline{S}_{\tau,c} \to S_\tau$ and let $\underline{s}'_0 \in \underline{S}^n_{\tau,c}$ map onto $\underline{s} \in \underline{S}_{\tau,c}$. Pick a connected étale neighbourhood of $\underline{s}$

$$\eta':\underline{T}(\underline{t}') \to \underline{S}_\tau$$

with $\eta'(\underline{t}') = \underline{s}$, such that there exists a morphism

$$\rho':\underline{T}(\underline{t}') \to \underline{H}_0(\underline{s})$$

with the property that

$$\eta'^*(\tilde{X}_\tau) \simeq \rho'^*(\tilde{X}_0(\underline{x}))$$

where as above $\pi_\tau:\tilde{X}_\tau \to \underline{S}_\tau$ is the restriction of π to $\underline{S}_\tau$ and $\pi_0(\underline{s}):\tilde{X}_0(\underline{s}) \to \underline{H}_0(\underline{s})$ is the restriction of $\pi(\underline{s})$ to $\underline{H}_0(\underline{s})$. Recall that, under our conditions, we may assume ρ' smooth. Take the pull-back of η' to $\underline{S}^n_{\tau,c}$ and get a Cartesian diagram

$$\begin{array}{ccccc}
\underline{H}_0(\underline{s}) & \xleftarrow{\rho'} & \underline{T}(\underline{t}') & \xrightarrow{\eta'} & \underline{S}_\tau \\
 & & \nu\uparrow & & \uparrow \\
 & & \underline{T}(\underline{t}) & \xrightarrow{\eta} & \underline{S}^n_{\tau,c}
\end{array}$$

Since $\underline{S}^n_{\tau,c}$ is integral, $\underline{T}(t)$ is also integral and ν is the normalization of its image in $\underline{T}(t')$. η is an étale morphism. Restricting to an open dense subset $\underline{S}'_{\tau,c}$ of $\underline{S}^n_{\tau,c}$ we may assume η is finite. Therefore we may as well assume $\underline{T}(\underline{t}) \xrightarrow{\eta} \underline{S}'_{\tau,c}$ is a Galois covering with Galois group G. Let $\underline{H}_0(\underline{s})_c$ be the reduced image of $\rho'\circ\nu$. Since $\underline{T}(\underline{t})$ is reduced, irreducible and connected, $\underline{H}_0(\underline{s})_c$ is a reduced component of $\underline{H}_0(\underline{s})$. Put $\underline{H}_0 = \underline{H}_0(\underline{s})_c$ and let $\rho_0:\underline{H}_1 \to \underline{H}_0$ be the normalization of $\underline{H}_0$. To simplify notations, we shall put $\underline{T} = \underline{T}(\underline{t})$, $\underline{S}' = \underline{S}'_{\tau,c}$. Consider now the diagram

$$\begin{array}{ccccc}
\underline{H}_0(\underline{s}) & \xleftarrow{\rho'} & \underline{T}(\underline{t}') & \xrightarrow{\eta'} & \underline{S}_\tau \\
\cup & & & & \\
\underline{H}_0 & & \uparrow\nu & & \Big\uparrow \\
\rho_0\uparrow & & & & \\
\underline{H}_1 & \xleftarrow{\rho} & \underline{T}' & & \\
 & & \uparrow & & \\
 & & \underline{T} & \xrightarrow{\eta} & \underline{S}'
\end{array}$$

where ρ is the pull-back of ρ'. Since $\underline{H}_1$ is normal and ρ is smooth we know by (EGA IV (7.3.8) that $\underline{T}'$ is normal, therefore $\underline{T}' \simeq \underline{T}$.

Let $\pi_0 : \tilde{X}_0 \to \underline{H}_0$ be the pull-back of $\pi_0(\underline{s})$ by $\underline{H}_0 \to \underline{H}_0(\underline{s})$, and let $\pi' : \tilde{X}' \to \underline{S}'$ be the pull-back of π_τ by $\underline{S}' \to \underline{S}_\tau$. Then we know (identifying $\underline{T}'$ and $\underline{T}$) that

$$\eta^*(\tilde{X}') \simeq (\rho_0 \circ \rho)^*(\tilde{X}_0).$$

Let $\pi : \tilde{X} \to \underline{T}$ be this common pull-back of π_0 and π'. We may assume $\underline{H}_i = \mathrm{Spec}(H_i)$, $i = 0, 1$, $\underline{T} = \mathrm{Spec}(T)$ and $\underline{S}' = \mathrm{Spec}(S')$. Then, since η is étale, we find

$$V(\pi') \subseteq \mathrm{Der}_k(S')$$

$$V(\pi) \subseteq \mathrm{Der}_k(T)$$

$$V(\pi') \otimes_{S'} T \simeq V(\pi)$$

To see this observe that $\mathrm{Der}_k(S', T) \simeq \mathrm{Der}_k(T, T)$ and therefore by (3.3) (iv), $V(\pi) = V(\pi', \eta)$. Moreover $V(\pi') \otimes_{S'} T \simeq V(\pi', \eta)$. The last statement follows from the fact that T is locally free of finite rank on S', coupled with (3.3) (i).
In particular, this implies

$$S'^V \to T^V$$

and for every $g \in \mathrm{Gal}(T/S')$ and $t \in T^V$, $g(t) \in T^V$. Thus T^V is stable under $G = \mathrm{Gal}(T/S')$, and obviously $(T^V)^G = S'^V$. We know already (see (3.16) (iii)) that

$$V(\pi_0'(\underline{t}')) \simeq \mathrm{Der}_{H_0(\underline{s})}(T(\underline{t}'))$$

where $\pi_0'(\underline{t}')$ is the pull-back of π_τ by η'. From this follows easily that

$$V(\pi) = \mathrm{Der}_{H_1}(T) = \mathrm{Der}_{T^V}(T).$$

In fact, since ρ' is smooth $\mathrm{Der}_{H_0(\underline{s})}(T(\underline{t}'))_{T(\underline{t}')} \otimes T \simeq \mathrm{Der}_{H_1}(T)$ $\subseteq V(\pi)$. Notice also that $V(\pi_0(\underline{s})) = 0$ therefore $V(\pi_0) = 0$ and so

$V(\rho_0^{\star}(\pi_0)) = 0$ since ρ_0 is the normalization map. This implies $H_1 \subseteq T^V$. As $V(\pi)$ has the same rank as $V(\pi_0'(\underline{t}'))$, namely the dimension of the fibers of ρ', $Der_{H_1}(T)$ and $V(\pi)$ has the same rank everywhere, therefore $Der_{H_1}(T) = V(\pi)$.

It follows that the codimension of ρ which is equal to the rank of $V(\pi)$ is equal to the rank of $Der_{T^V}(T)$. However, rank $Der_{T^V}(T) = tr.deg.(K(T)/K(T^V))$.

Consequently $K(H_1) \subseteq K(T^V)$ has trancendence degree 0, and is therefore algebraic. We shall show that $K(H_1) \subseteq K(T^V)$ is a finite extension.

In fact, if $K(H_1) \subseteq K_2$ is a finite subextension and H_2 is the integral closure of H_1 in K_2, then $\rho_1 : \underline{H}_2 \to \underline{H}_1$ is finite. Since T is normal, T^V is also normal, therefore $H_2 \subseteq T^V$. Given $\underline{s}' \in \underline{S}'$ $\eta^{-1}(\underline{s}') = \{\underline{t}_1, \ldots, \underline{t}_r\}$ and let $\underline{h}_i = \rho_2(\underline{t}_i)$, $i = 1, \ldots, r$, where $\rho_2 : \underline{T} \to \underline{H}_2$ is the obvious morphism. Since all fibers of $\rho : \underline{T} \to \underline{H}_1$ are smooth, the points $\underline{h}_1, \ldots, \underline{h}_r$ must be unramified for ρ_1. It follows from (E.G.A. IV.(8.4.9)) that ρ_1 is étale at $\underline{h}_j$, $j = 1, \ldots, r$. But then $\rho_2 : \underline{T} \to \underline{H}_2$ is smooth at the points $\underline{h}_j \in \underline{H}_2$, $j = 1, \ldots, r$, since the composition $\underline{T} \xrightarrow{\rho_2} \underline{H}_2 \xrightarrow{\rho_1} \underline{H}_1$ is smooth. Therefore im ρ_2 contains an open set $\underline{U}_2 \subseteq \underline{H}_2$ containing $\underline{h}_1, \ldots, \underline{h}_r$.
In particular $\underline{U}_2$ will contain the generic fiber of ρ_1.

Now suppose there exist an infinite chain of finite sub-extensions $K(H_1) \subseteq K_2 \subseteq K_3 \subseteq \ldots \subseteq K(T^V)$. Let H_i be the integral closure of H_1 in K_i, then ρ splits up into an infinite composition

$$\underline{T} \to \underline{T}^V \to \ldots \to \underline{H}_{i+1} \to \underline{H}_i \to \ldots \to \underline{H}_1 .$$

The generic fiber of ρ would then have an infinite number of connected components, which is nonsense.

Therefore $K(H_1) \subseteq K(T^V)$ is a finite extension. Put $K_2 = K(T^V)$ and let $\underline{U}_2 \subseteq \underline{H}_2$ be as above. Since $\underline{H}_2$ is affine, there exists an $m \in H_2$ such that the open set $D(m)$ is contained in $\underline{U}_2$ and contains $\underline{h}_1, \ldots, \underline{h}_r$. In particular $m(\underline{h}_i) \neq 0$, $i = 1, \ldots, r$. Let $N(m) = \prod_{g \in G} g(m)$ be the norm of m. Obviously $N(m) \in S'^V$ and since the elements of G

permutes the points $\underline{t}_1,\ldots,\underline{t}_r$ we have $N(m)(\underline{t}_i) \neq 0$, $i = 1,\ldots,r$, and thus $N(m)(\underline{s}') \neq 0$. Therefore $\underline{h}_1,\ldots,\underline{h}_r \in D(N(m)) \subseteq D(m) \subseteq U_2$. By construction the restriction of ρ_2 is surjective on $D(N(m))$. Since $H_{2\{N(m)\}} \subseteq T^V_{\{N(m)\}} \subseteq T_{\{N(m)\}}$ and $D(N(m)) = \mathrm{Spec}(H_{2\{N(m)\}})$ it follows that

$$\mathrm{Spec}(T^V_{\{N(m)\}}) \to \mathrm{Spec}(H_{2\{N(m)\}})$$

is onto.

As $H_{2\{N(m)\}}$ obviously is normal, it follows from (3.19) that $H_{2\{N(m)\}} = T^V_{\{N(m)\}}$, in particular $T^V_{\{N(m)\}}$ is of finite type. From (3.21) we conclude that G operates on $T^V_{\{N(m)\}}$ without fixpoints. Therefore

$$S^V_{\{N(m)\}} = (T^V_{\{N(m)\}})^G \subseteq T^V_{\{N(m)\}}$$

is étale and $S^V_{\{N(m)\}}$ is of finite type. Summing up, what we have proved so far is:

Given any $\underline{s}'_0 \in \underline{S}'_{\tau,c}$ there exists an affine neighbourhood $\underline{S}' \subseteq \underline{S}'_{\tau,c}$ of $\underline{s}'_0$ such that for every $\underline{s}' \in \underline{S}'$ there exists an element $N(\underline{s}') \in S'^V$ such that $\underline{s}' \in D(N(\underline{s}'))$ and $S'^V_{\{N(\underline{s}')\}}$ is of finite type.

For every $\underline{s}' \in \underline{S}'$ take $\underline{N}(\underline{s}') = \mathrm{Spec}(S'^V_{\{N(\underline{s}')\}})$ and glue. The glueing is straightforward, and we obtain a scheme of finite type $\underline{N}_{\tau,c}$, the quotient of $\underline{S}'_{\tau,c}$ by $V(\pi')$. Since locally the morphism $\rho_{\tau,c}: \underline{S}'_{\tau,c} \to \underline{N}_{\tau,c}$ sits in the following diagram

$$\begin{array}{ccc} \underline{T}_{\{N(m)\}} & \xrightarrow{\eta} & \underline{S}'_{\tau,c} \\ \rho \downarrow & & \downarrow \rho_{\tau,c} \\ \underline{H}_{2\{N(m)\}} & \xrightarrow{\eta_2} & \underline{N}_{\tau,c} \end{array}$$

where ρ is smooth, η and η_2 are étale, it follows that $\rho_{\tau,c}$ is smooth. Moreover, since by (3.22) the descent-datum on $\eta^*(\pi')$ descend to a descent-datum on $(\rho_0 \circ \rho_1)^*(\pi_0)$ for η_2, this family descends to $\underline{S}'^V_{\{N(m)\}}$. Glue these families and obtain a family $\mu_{\tau,c}: \mathfrak{X}_{\tau,c} \to \underline{N}_{\tau,c}$, which must pull back by $\rho_{\tau,c} \circ \eta$ to $\eta^*\pi'$. The rest is obvious.

Q.E.D.

Notice that if there exists a coarse moduli space for the objects of the family π_τ, then, locally, it must be a discrete quotient of $\underline{M}_\tau$. Therefore the dimension of $\underline{N}_{\tau,c}$ is the correct "number of moduli" in the sense of Riemann.

Definition (3.25). We shall refer to the collection $\underline{M} = \{\underline{M}_\tau\}$ as the local moduli suite of X.

The strict modality of X is the integer

$$sm(X) = \dim \underline{H}_0.$$

By the τ-modality of X we shall mean the integer

$$\tau m(X) = \max\{\dim \underline{N}_{\tau,c} \mid c\}.$$

The modality of X is defined to be

$$m(X) = \max\{\tau m(X) \mid 0 \leq \tau \leq \tau^1\}.$$

Suppose, as above, that $\underline{H}$ is non-singular, and suppose the generic fiber of $\pi : \tilde{X} \to \underline{H}$ is rigid. Then by (3.10) $\tau(X) = \dim \underline{H}$ is equal to $\dim_k (V \otimes_H k)$. Therefore the inclusion $V \subseteq \mathrm{Der}_k(H)$ induces an H-linear map $i : \overset{\tau}{\wedge} V \to \overset{\tau}{\wedge} \mathrm{Der}_k(H) = H$, the image of which is a principal ideal.

Definition (3.26). The image Δ of i, a principal ideal generated by the determinant of V, is called the discriminant of π. The corresponding hypersurface will be denoted by $\underline{\Delta}$.

Remark (3.27). By definition $\underline{\Delta} = \bigcup_{\tau > 0} \underline{S}_\tau$.

§4 APPLICATIONS TO ISOLATED HYPERSURFACE SINGULARITIES

Introduction. Up to now we have been studying the problem of local moduli in as general a setting as possible.

In this and the following paragraphs we shall restrict our attention to hypersurface singularities in the algebroid sense.

If $X = \mathrm{Spec}(k[x_1,\ldots,x_n]/(f))$ is a hypersurface with isolated singularities, then as we will see, X satisfies the conditions (A_1). and (V') but not necessarily (A_2). Therefore we cannot invoke the theorems (3.18) and (3.24). If, however, we restrict our attention to the completions $X = \mathrm{Spf}(k[[x_1,\ldots,x_n]]/(f))$, we have to be a little careful with respect to which category we are working in, and we have to change our cohomology slightly, but otherwise we find ourselves in the situation of §3.

The basic fact is, of course, that in this algebroid situation Artins approximation theorem applies so that (A_2) holds.

Thus, in particular, there is, in the category of algebraic spaces, a local moduli suite $\{\underline{M}_\tau\}$, see (3.18), and (3.24) holds. The problem of actual computation of the components $\underline{M}_\tau$ seems, however at this point outside of our reach.

Now, for hypersurface singularities, we have at our disposal the algebro-topological invariant μ, the Milnor number, and we may restrict our attention to the μ-constant stratum $\underline{H}_\mu$ of $\underline{H}$, and to the corresponding μ-constant substratum $\underline{M}_{\mu\tau}$ of $\underline{M}_\tau$. This amounts to fixing the topological type of the singularity. The structure of these substrata is however still quite complicated, even in the quasihomogenous case. One problem is that $\underline{S}_\tau$ may contain one component inside $\underline{H}_\mu$, and another component intersecting the first one but not included in $\underline{H}_\mu$, see §5 for examples.

The main result, (4.5), is an "explicit" calculation of the kernel of the Kodaira-Spencer map of π_μ. This result is the basic tool in our study of the curve case in §5 and §6.

From now on X will be a hypersurface f, i.e. $X = \mathrm{Spec}(A)$, $A = k[x_1,\dots,x_n]/(f)$ where $f = f(x_1,\dots,x_n) \in k[x_1,\dots,x_n]$. Suppose f has only isolated singularities, then X satisfies the conditions (V') of §3. Moreover

$$A^1(k,X;O_X) = k[x_1,\dots,x_n]/(f,\frac{\partial f}{\partial x_1},\dots,\frac{\partial f}{\partial x_n}).$$

Pick a base for $A^1(k,X;O_X)$ represented by $\{\lambda_i\}_{i=1}^{\tau^1}$, $\lambda_i \in k[\underline{x}]$, where as above,

$$\tau^1 = \tau(f) = \dim_k A^1(k,X;O_X).$$

Put, $H = k[t_1,\dots,t_{\tau^1}]$, $F = f + \sum_{i=1}^{\tau^1} \lambda_i t_i \in H[\underline{x}]$ and $\tilde{A} = H[\underline{x}]/(F)$. Then $\tilde{X} = \mathrm{Spec}(\tilde{A}) \to \underline{H} = \mathrm{Spec}(H)$ is the family with which we shall have to work.

<u>Remark (4.1)</u>. Since (A_2) of §3 is not satisfied for affine schemes in general, we cannot use the theorems (3.18) and (3.24). But notice that the prorepresenting substratum $\underline{H}_0$ of $\underline{H}$ still exists.

<u>Proposition (4.2)</u>. The Kodaira-Spencer map

$$g: \mathrm{Der}_k(H) \to A^1(H,\tilde{A};\tilde{A})$$

is given by

$$g(\frac{\partial}{\partial t_i}) = \text{class of } \frac{\partial F}{\partial t_i} \text{ in } H[x_1,\dots,x_n]/(F,\frac{\partial F}{\partial x_1},\dots,\frac{\partial F}{\partial x_n}).$$

<u>Proof</u>. By definition of g, it is the obstruction for lifting $1_{\tilde{A}} \in \mathrm{Aut}_H(\tilde{A})$ to an isomorphism $\phi: i_0^*(\tilde{A}) \to i_1^*(\tilde{A})$ where $i_k: H \to H\otimes H/I^2$ $k = 1,2$ and $I = \ker\{H\otimes H \to H\}$ are defined as the ν_k's in §3.

Now $H\otimes H/I^2 = k[\underline{t},\underline{u}]/(t_i-u_i)^2$, $i_0^*(\tilde{A}) = k[\underline{t},\underline{u}][\underline{x}]/F_t$, $i_1^*(\tilde{A}) = k[\underline{t},\underline{u}][\underline{x}]/F_u$, where $F_t = f+\Sigma\lambda_i t_i$, $F_u = f+\Sigma\lambda_i u_i$. Obviously this obstruction is simply given by the difference $(F_t - F_u) = \Sigma\lambda_i(t_i-u_i)$ in $A^1(H,\tilde{A};\tilde{A}\otimes I/I^2) = H[\underline{x}]/(F,\frac{\partial F}{\partial x_1},\dots,\frac{\partial F}{\partial x_n}) \otimes_H I/I^2$. But then $g = \Sigma(\text{class of } \lambda_i)\cdot dt_i$.

Q.E.D

Now, in this § we shall concentrate on the deformations of the completion $\hat{A}_{(\underline{x})} = k[[\underline{x}]]/(f)$ as a complete augmented k-algebra, see [Z]. (We shall refer to $\hat{A}_{(\underline{x})}$ or to f as the hypersurface singularity). By a deformation of $\hat{A}_{(\underline{x})} \to k$, to an augmented k- algebra $R \to k$ we shall understand any commutative diagram

$$\begin{array}{ccccc} R & \to & A_R & \overset{\rho}{\to} & R \\ \downarrow & & \downarrow & & \downarrow \\ k & \to & \hat{A}_{(\underline{x})} & \to & k \end{array}$$

where $\rho: A_R \to R$ is a flat augmented R-algebra such that $A_R = \varprojlim_n A_R/(\ker\rho)^n$, and where $A_R \otimes_R k = \hat{A}_{(\underline{x})}$.

We may develop a cohomology theory for this case, with a corresponding obstruction theory, and we may prove the existence of a formal moduli. Moreover, for any flat augmented R-algebra $\rho: A_R \to R$ such that $A_R = \varprojlim_n A_R/(\ker\rho)^n$, there exists a Kodaira-Spencer map just as above. The cohomology is given by

$$A^0(k, \hat{A}_{(\underline{x})} \to k;\ \hat{A}_{(\underline{x})}) = \mathrm{Der}_k^c(\hat{A}_{(\underline{x})})$$

$$A^1(k, \hat{A}_{(\underline{x})} \to k; \hat{A}_{(\underline{x})}) = H^1(k, \hat{A}_{(x)}; (\underline{x})\hat{A}_{(x)}) = (\underline{x})/(\underline{x})(\frac{\partial f}{\partial x_1}, \dots, \frac{\partial f}{\partial x_n}) + (f)$$

$$= \ker\{k[[\underline{x}]]/(f, \frac{\partial f}{\partial x_1}, \dots, \frac{\partial f}{\partial x_n}) \to k\} \oplus$$

$$\mathrm{coker}\{\mathrm{Der}_k(k[[x]]/(f)) \to \mathrm{Der}_k(k[[x]]/(f), k)\}$$

$$A^i(k, \hat{A}_{(\underline{x})} \to k;\ \hat{A}_{(\underline{x})}) = 0 \quad \text{for} \quad i \geq 2$$

$$A^1(R, A_R \to R; A_R) = H^1(R, A_R; \ker\rho).$$

Let us put, as customary, $\tau(f) = \dim_k k[[\underline{x}]]/(f, \frac{\partial f}{\partial x_1}, \dots, \frac{\partial f}{\partial x_n})$, the Tjurina number of the singularity. Notice that if $f \in (\underline{x})^2$ $\dim A^1 = \tau(f)+n-1$. Pick any basis $\{t_1^*, \dots, t_s^*\}$ for $A^1(k, \hat{A}_{(x)} \to k; \hat{A}_{(x)})$, then the formal moduli of $\hat{A}_{(x)} \to k$ is $\hat{H.} = k[[t_1, \dots, t_s]]$. The formal versal family $X.^{\wedge} = H.[[\underline{x}]]^{\wedge}/(F.)$, is given by $F. = f + \sum_{i=1}^{s} g_i \cdot t_i$, where g_i are representatives in $(\underline{x}) \cdot k[[\underline{x}]]$ of t_i^*, and where $H.[[x]]^{\wedge}$ means completion in the $(\underline{t})$-adic topology. From now on we shall always pick a basis $\{t_1^*, \dots, t_s^*\}$ for which we may choose the representatives g_i to be monomials in $k[[\underline{x}]]$. Put $g_i = \underline{x}^{\underline{\alpha}_i}$, $i = 1, \dots, s$, then

$F. = f + \sum_{i=1}^{s} t_i \underline{x}^{\underline{\alpha}_i}$. Now let us consider the complete augmented H.-algebra $H.[[\underline{x}]]/(F.) \to H.$ where $H. = k[t_1,\ldots,t_s]$ and put $\underline{H}. = \mathrm{Spec}(H.)$, $\tilde{X}. = \mathrm{Spf}(H.[[\underline{x}]]/(F.))$. Then

$$\underline{H}. \to \tilde{X}. \to \underline{H}.$$

is a versal deformation of the pointed pro-scheme $\mathrm{Spec}(k) \to \mathrm{Spf}(\hat{A}_{(x)}) \to \mathrm{Spec}(k)$.

The corresponding Kodaira-Spencer map

$$g: \mathrm{Der}_k(H.) \to A^1(H., H.[[\underline{x}]]/(F.) \to H.; H.[[\underline{x}]]/(F.))$$

is given by

$$g\left(\frac{\partial}{\partial t_i}\right) = \text{class of } \frac{\partial F.}{\partial t_i} = \text{class of } \underline{x}^{\underline{\alpha}_i}, \text{ in}$$

$$(\underline{x}) \cdot H.[[\underline{x}]]/((F.) + (\frac{\partial F.}{\partial x_1},\ldots,\frac{\partial F.}{\partial x_n}) \cdot (\underline{x})).$$

Now, enumerate the flattening stratification of $V = \ker g$ such that $\tau(F.(\underline{t})) = \tau$ for $\underline{t} \in \underline{S}_\tau$. This is possible since $\dim A^1 = \dim_k (\underline{x})/(\underline{x})(\frac{\partial f}{\partial x_1},\ldots,\frac{\partial f}{\partial x_n}) + (f) = \tau(f)+n-1$.
Copying the proof of (3.18) and (3.24) we then prove the following result.

<u>Theorem (4.3)</u>. Let $f \in k[\underline{x}]$ be an isolated singularity and put $\tau_0 = \tau(f)$. Then there exists, in the category of algebraic spaces, a local moduli suite $\{\underline{M}_\tau\}_{\tau=0}^{\tau_0}$ for f. Moreover, there exists a finite collection of families of augmented algebraic schemes,

$$\mu_{\tau,c}: \mathcal{X}_{\tau,c} \to \underline{N}_{\tau,c} \qquad 0 \leq \tau \leq \tau_0$$

such that

(i) $\underline{N}_{\tau,c}$ is the quotient of an open dense subscheme of $\underline{S}^n_{\tau,c}$ by the Kodaira-Spencer kernel V (see (3.24)).

(ii) There exists a quasifinite dominant morphism $\underline{N}_{\tau,c} \to \underline{M}_\tau$ such that $\mu_{\tau,c}$ is the pull-back of $\bar{\pi}_\tau: \bar{X}_\tau \to \underline{M}_\tau$.

There seems at present to be little hope getting much further in the study of the local moduli suite $\underline{M} = \{\underline{M}_\tau\}$ for general hypersurface singularities.

However, if we restrict ourselves to quasihomogenous singularities f, and if we fix the Milnor number we are in a much better situation. The rest of this § is therefore devoted to the study of <u>topologically constant deformations of isolated hypersurface singularities f, where f is quasihomogenous of weights</u> $w_1,\dots,w_n$ <u>and degree 1.</u>

Recall the definition of the Milnor number of an isolated hypersurface singularity f,

$$\mu(f) = \dim_k k[[x_1,\dots,x_n]]/(\frac{\partial f}{\partial x_1},\dots,\frac{\partial f}{\partial x_n}),$$

and let from now on $\underline{H}_\mu \subseteq \underline{H}.$ denote the closed substratum of $\underline{H}.$ where μ is constant and equal to $\mu = \mu(f)$.

Assume that:

$$f = \sum_{i=1}^{n} x_i^{a_i}, \quad w_i = \frac{1}{a_i}, \; a_i \geq 2, \quad i = 1,\dots,n.$$

In this case $A^1(k,k[[\underline{x}]]/(f) \to k,k[[\underline{x}]]/(f)) =$
$(\underline{x})\cdot k[[\underline{x}]]/(\frac{\partial f}{\partial x_1},\dots,\frac{\partial f}{\partial x_n})\cdot(\underline{x})$ has a monomial basis $\{\underline{x}^{\underline{\alpha}} | \underline{\alpha}\in I.\}$ where, if $I = \{(\alpha_1,\dots,\alpha_n)\in Z_+^n | 0 \leq \alpha_i \leq a_i - 2\}$, then
$I. = I\cup\{(0,\dots,a_i-1,0\dots) | i=1,\dots.n\} \; \{(0,\dots,0)\}$. Put
$I_\mu = \{\underline{\alpha}\in I. \,|\, |\underline{\alpha}| \geq 1\}$ where $|\underline{\alpha}| = \sum_{i=1}^{n} w_i\alpha_i$, then

$$H. = k[t_{\underline{\alpha}}]_{\underline{\alpha}\in I.}$$

$$H_\mu = H/(t_{\underline{\alpha}} | \underline{\alpha}\in I.\setminus I_\mu) = k[t_{\underline{\alpha}}]_{\underline{\alpha}\in I_\mu}.$$

Moreover H. is a graded k-algebra with $\deg t_{\underline{\alpha}} = 1 - |\underline{\alpha}|$. In particular H_μ is therefore a negatively graded k-algebra.

The versal family F. restricted to $\mathrm{Spec}(H_\mu) = \underline{H}_\mu \subseteq \underline{H}.$, is then given as

$$F_\mu = f + \sum_{\underline{\alpha}\in I_\mu} t_{\underline{\alpha}}\underline{x}^{\underline{\alpha}}, \text{ with } \deg F_\mu = 1.$$

Notice that $H_\mu[[\underline{x}]]/(\frac{\partial F_\mu}{\partial x_i})$ is a graded k-algebra, with $\deg x_i = w_i$, and locally free as an H_μ-module, with basis $\{\underline{x}^{\underline{\alpha}}\}_{\underline{\alpha}\in I}$.

We shall have to compute the kernel $V_\mu = V(\pi_\mu)$ of the Kodaira-Spencer map in the algebroid sense. V_μ coincide with the kernel of following "Kodaira-Spencer" map

$$g_\mu : \mathrm{Der}(H_\mu) \to H_\mu[[\underline{x}]]/(F_\mu, \frac{\partial F_\mu}{\partial x_1}, \ldots, \frac{\partial F_\mu}{\partial x_n})$$

given by $g_\mu(\frac{\partial}{\partial t_{\underline{\alpha}}})$ = class of $\frac{\partial F_\mu}{\partial t_{\underline{\alpha}}}$, see (4.2). Thus we want to compute $\ker g_\mu$. But before this, let's consider the Lie algebra $\mathrm{Der}_k(k[[\underline{x}]]/(f))$. Since f is quasihomogenous, there is one special derivation $D_0 \in \mathrm{Der}_k(k[[x]]/(f))$, defined by

$$D_0(x_i) = w_i \cdot x_i, \; i = 1, \ldots, n.$$

Moreover, for every $i,j = 1,\ldots,n$, there is a derivation $E_{ij} \in \mathrm{Der}_k(k[[\underline{x}]]/(f))$ defined by

$$E_{ij}(x_k) = \begin{cases} 0, & k \neq i,j \\ \frac{\partial f}{\partial x_j}, & k = i \\ -\frac{\partial f}{\partial x_i}, & k = j \end{cases}$$

Clearly $E_{ij} \in \mathrm{Der}_\pi$, for all $i,j = 1,\ldots,n$, where $\mathrm{Der}_\pi \subseteq \mathrm{Der}_k(k[[\underline{x}]]/(f))$, as in §3, is the Lie ideal consisting of those derivations that may be lifted everywhere. Given any $D \in \mathrm{Der}(k[[\underline{x}]]/(f))$ there exist representatives $\xi_i \in k[[\underline{x}]]$ of $D(x_i)$ such that

$$\sum_{i=1}^{n} \frac{\partial f}{\partial x_i} \cdot \xi_i = q \cdot f$$

for some $q \in k[[\underline{x}]]$. Recalling the Euler identity $\sum_{i=1}^{n} w_i x_i \frac{\partial f}{\partial x_i} = f$ we find the relation

$$\sum_{i=1}^{n} \frac{\partial f}{\partial x_i}(\xi_i - q\cdot w_i x_i) = 0.$$

Since f has an isolated singularity at the origin, the only relations among the $\frac{\partial f}{\partial x_i}$'s are the trivial ones.

It follows that $D - q\cdot D_0 = \sum r_{ij} E_{ij}$, $r_{ij} \in k[[\underline{x}]]$, i.e.

$$D - q\cdot D_0 \in \mathrm{Der}_\pi$$

thus the map

$$\phi : k[[\underline{x}]]/(\frac{\partial f}{\partial x_1}, \ldots, \frac{\partial f}{\partial x_n}) \to \mathrm{Der}_k(k[[x]]/(f))/\mathrm{Der}_\pi$$

defined by $\phi(q) = q\cdot D_0$, is surjective.

For every $\underline{\alpha} \in I.$, let's denote by $D_{\underline{\alpha}}$ the image by ϕ of $\underline{x}^{\underline{\alpha}}$, i.e.

$$D_{\underline{\alpha}} = \underline{x}^{\underline{\alpha}} \cdot D_0.$$

One may easily prove the following

<u>Lemma (4.4)</u>. With the notations above, we have

(i) $[D_{\underline{\alpha}}, D_{\underline{\beta}}] = (|\underline{\beta}| - |\underline{\alpha}|) D_{\underline{\alpha}+\underline{\beta}}$

(ii) $L = \mathrm{Der}_k(k[[\underline{x}]]/(f))/\mathrm{Der}_\pi$ is solvable

(iii) $[L,L] = L_0$ is nilpotent and L/L_0 is generated by $D_{\underline{o}}$.

It is not difficult to see that ϕ is an isomorphism. In fact, for any isolated hypersurface singularity f we may consider the multiplication by f as a map

$$f^{\cdot} : k[[\underline{x}]]/(\frac{\partial f}{\partial x_1}, \ldots, \frac{\partial f}{\partial x_n}) \to k[[\underline{x}]]/(\frac{\partial f}{\partial x_1}, \ldots, \frac{\partial f}{\partial x_n})$$

Put

$$L(f) = \mathrm{Der}_k(k[[\underline{x}]]/(f))/\mathrm{Der}_\pi$$

then we easily prove that there is an isomorphism

$$i : L(f) \simeq \ker f^{\cdot}$$

which we may describe as follows. Let $d \in L(f)$ be represented by

$D \in \mathrm{Der}(k[[\underline{x}]])$. Then there exists a $q \in k[[\underline{x}]]$ such that

$$D(f) = \sum \frac{\partial f}{\partial x_i} D(x_i) = q \cdot f$$

Obviously q represents an element $\bar{q} \in \ker f^{\cdot}$, and

$$i(d) = \bar{q}$$

Now, returning to the Kodaira-Spencer map, we observe that g_μ may be factorized into

$$g_0 : \mathrm{Der}_k(H_\mu) \to H_\mu[[\underline{x}]]/(\frac{\partial F_\mu}{\partial x_1}, \dots, \frac{\partial F_\mu}{\partial x_n})$$

and the natural map

$$H_\mu[[\underline{x}]]/(\frac{\partial F_\mu}{\partial x_1}, \dots, \frac{\partial F_\mu}{\partial x_n}) \to H_\mu[[\underline{x}]]/(F_\mu, \frac{\partial F_\mu}{\partial x_1}, \dots, \frac{\partial F_\mu}{\partial x_n})$$

where $g_0(\frac{\partial}{\partial t_{\underline{\alpha}}}) = \underline{x}^{\underline{\alpha}}$ for $\underline{\alpha} \in I_\mu$.

Since $H_\mu[[\underline{x}]]/(\frac{\partial F_\mu}{\partial x_1}, \dots, \frac{\partial F_\mu}{\partial x_n})$ is locally free with basis $\{\underline{x}^{\underline{\alpha}}\}_{\underline{\alpha} \in I}$, g_0 is clearly injective. Moreover let $\{\underline{x}^{\underline{\alpha}*}\}_{\underline{\alpha} \in I}$ be the dual basis, then $\underline{x}^{\underline{\alpha}*}$, restricted to $\mathrm{Der}_k(H_\mu)$, may be identified with the map $dt_{\underline{\alpha}} : \mathrm{Der}_k(H_\mu) \to H_\mu$.

Consider

$$E = \sum_{i=1}^{n} w_i x_i \frac{\partial F_\mu}{\partial x_i} - F_\mu = \sum_{\underline{\alpha} \in I_\mu} (|\underline{\alpha}| - 1) \underline{x}^{\underline{\alpha}} t_{\underline{\alpha}}$$

as an element of $H_\mu[[\underline{x}]]/(\frac{\partial F_\mu}{\partial x_1}, \dots, \frac{\partial F_\mu}{\partial x_n})$. For every $\underline{\alpha} \in I$, we may write

$$E \cdot \underline{x}^{\underline{\alpha}} = \sum_{\underline{\beta} \in I_\mu} k_{\underline{\alpha}\underline{\beta}} \underline{x}^{\underline{\beta}}, \quad k_{\underline{\alpha}\underline{\beta}} \in H_\mu$$

where $k_{\underline{\alpha}\underline{\beta}}$ are homogeneous of degree $1 + |\underline{\alpha}| - |\underline{\beta}|$.

<u>Proposition (4.5)</u>. With the notations above,

(i) V_μ is a graded Lie algebra locally generated as H_μ-module by the elements

$$\delta_{\underline{\alpha}} = \sum_{\underline{\beta} \in I_\mu} k_{\underline{\alpha}\underline{\beta}} \frac{\partial}{\partial t_{\underline{\beta}}}, \text{ with } \underline{\alpha} \in I \text{ and } \deg \delta_{\underline{\alpha}} = |\underline{\alpha}|$$

(ii) The canonical morphism

$$\lambda_\mu : V_\mu \to \mathrm{Der}(k[[\underline{x}]]/(f))/\mathrm{Der}_\mu$$

where Der_μ is the Lie ideal of those derivations that may be lifted to H_μ, is given by

$$\lambda_\mu(\delta_{\underline{\alpha}}) = -D_{\underline{\alpha}}.$$

(iii) $[\delta_{\underline{\alpha}}, \delta_{\underline{\beta}}] = (|\underline{\alpha}|-|\underline{\beta}|)\delta_{\underline{\alpha}+\underline{\beta}} + \sum_{|\underline{\gamma}|>|\underline{\alpha}|+|\underline{\beta}|} h_{\underline{\alpha},\underline{\beta},\underline{\gamma}}\delta_{\underline{\gamma}}$ where $h_{\underline{\alpha},\underline{\beta},\underline{\gamma}} \in (t_{\underline{\alpha}})_{\underline{\alpha}\in I_\mu} \subseteq H_\mu$.

Proof. Obviously $\underline{x}^{\underline{\alpha}}\cdot E$ is zero in $H_\mu[[\underline{x}]]/(F_\mu, \frac{\partial F_\mu}{\partial \underline{x}})$ therefore $\delta_{\underline{\alpha}} \in V_\mu$. Conversely, suppose $\delta = \sum_{\underline{\beta}\in I_\mu} h_{\underline{\beta}} \frac{\partial}{\partial t_{\underline{\beta}}} \in V_\mu$, then $g_0(\delta) =$ $\sum_{\underline{\beta}\in I_\mu} h_{\underline{\beta}}\underline{x}^{\underline{\beta}} = \sum_{i=1}^{n} p_i \frac{\partial F_\mu}{\partial x_i} - pF_\mu = \sum_{i=1}^{n}(p_i - w_i x_i p)\frac{\partial F_\mu}{\partial x_i} + pE$. Now $pE = \sum_{\underline{\alpha}\in I} k_{\underline{\alpha}}\underline{x}^{\underline{\alpha}}E$, $k_{\underline{\alpha}} \in H_\mu$. In $H_\mu[[\underline{x}]]/(\frac{\partial F_\mu}{\partial x_1}, \ldots, \frac{\partial F_\mu}{\partial x_n})$ we have therefore

$$\sum_{\underline{\beta}\in I_\mu} h_{\underline{\beta}}\underline{x}^{\underline{\beta}} = \sum_{\underline{\alpha}\in I} k_{\underline{\alpha}} \sum_{\underline{\beta}\in I_\mu} k_{\underline{\alpha\beta}}\underline{x}^{\underline{\beta}}.$$

Since $\{\underline{x}^{\underline{\beta}}\}_{\underline{\beta}\in I_\mu}$ is part of a basis, it follows

$$\sum_{\underline{\beta}\in I_\mu} h_{\underline{\beta}} \frac{\partial}{\partial t_{\underline{\beta}}} = \sum_{\underline{\alpha}\in I} k_{\underline{\alpha}}\delta_{\underline{\alpha}}$$

proving (i).

(ii) Let $\delta_\alpha \in V_\mu$, then by (2.6), there exists a derivation $E_{\underline{\alpha}} \in \mathrm{Der}_k(H_\mu[[\underline{x}]]/(F_\mu))$ such that for all $k \in H_\mu$, $E_{\underline{\alpha}}(k\cdot x_i) = kE_{\underline{\alpha}}(x_i) +$ $\delta_\alpha(k)\cdot x_i$, $i = 1,\ldots,n$. Moreover $\lambda(\delta_{\underline{\alpha}})$ is the reduction of $E_{\underline{\alpha}}$. Now since $E_{\underline{\alpha}}$ is a k-derivation we must have $\sum_{i=1}^{n} \frac{\partial F_\mu}{\partial x_i}E_{\underline{\alpha}}(x_i) + \sum_{\underline{\beta}\in I_\mu} \frac{\partial F_\mu}{\partial t_{\underline{\beta}}}\delta_{\underline{\alpha}}(t_{\underline{\beta}})$ $= 0$ in $H_\mu[[\underline{x}]]/(F_\mu)$. On the other hand, put $E_{\underline{\alpha}}(x_i) = -\underline{x}^{\alpha} w_i x_i$, then

$$\sum_{i=1}^{n} \frac{\partial F_\mu}{\partial x_i} E_{\underline{\alpha}}(x_i) + \sum_{\underline{\beta} \in I_\mu} \underline{x}^{\underline{\beta}} k_{\underline{\alpha\beta}} = \sum_{i=1}^{n} \frac{\partial F_\mu}{\partial x_i} E_{\underline{\alpha}}(x_i) + \underline{x}^{\underline{\alpha}} \cdot E$$

is zero modulo (F_μ). Consequently we obtain a k-derivation $E_{\underline{\alpha}}$ with the properties we want, and for all $i = 1,\dots,n$ $\lambda(\delta_{\underline{\alpha}})(x_i) = E_{\underline{\alpha}}(x_i) = -\underline{x}^{\underline{\alpha}} w_i x_i$, i.e. $\lambda(\delta_\alpha) = -D_{\underline{\alpha}}$.

(iii) now follows from (4.4), (i) and (ii).

Q.E.D.

Before we continue the study of V_μ, let us pause, recalling the action ρ of $\mathrm{Der}(k[[\underline{x}]]/(f))$ on $A^1(k,\underline{k}[[x]]/(f)\to k;k[[\underline{x}]]/(f))$, see (2.12).

Corollary (4.6). With the notations above, we have:

(i) The action ρ of $\mathrm{Der}(k[[\underline{x}]]/(f))/\mathrm{Der}_\pi$ on $A^1(k,\underline{k}[[x]]/(f)\to k;k[[\underline{x}]]/(f))$ is given by

$$\rho(D_{\underline{\alpha}})(\underline{x}^{\underline{\beta}}) = (|\underline{\beta}|-1)\underline{x}^{\underline{\alpha}+\underline{\beta}}, \quad \underline{\alpha},\underline{\beta} \in I.$$

(ii) The corresponding action of V_μ on H_μ has a linear part given by: let $\underline{\gamma} \in I_\mu$, $\underline{\alpha} \in I$, then

$$\delta_{\underline{\alpha}}(t_{\underline{\gamma}}) \equiv \begin{cases} (|\underline{\gamma}-\underline{\alpha}|-1)t_{\underline{\gamma}-\underline{\alpha}} \ (\mathrm{mod}(\underline{t})^2) & \text{if } \underline{\gamma}-\underline{\alpha} \in I_\mu \\ 0 \ (\mathrm{mod}(\underline{t})^2) & \text{otherwise} \end{cases}$$

(iii) Consider the action ρ_0 of $\mathrm{Der}(k[[\underline{x}]]/(f))/\mathrm{Der}_\mu$ on the tangent space of H_μ at 0 given in terms of

$$\rho_0(D_{\underline{\alpha}})(\underline{x}^{\underline{\beta}}) = \begin{cases} 0 & \text{if } |\underline{\beta}| = 1 \\ \underline{x}^{\underline{\alpha}+\underline{\beta}} & \text{if } |\underline{\beta}| \neq 1 \end{cases}$$

then the dimensions of the maximal orbits of ρ and ρ_0 are equal.

Proof. (i) and (ii) are just variations on (2.12). To prove (iii) it suffices to see that the dimension of the orbit of $\sum t_{\underline{\beta}} \underline{x}^{\underline{\beta}}$ under ρ is the same as the dimension of the orbit of $\sum t_{\underline{\beta}}(|\underline{\beta}|-1)\underline{x}^{\underline{\beta}}$ under ρ_0.

Q.E.D.

Denote by $K = K(\underline{t})$ the matrix $(k_{\underline{\alpha\beta}})_{\underline{\alpha}\in I,\underline{\beta}\in I_\mu}$. Recall the notations from (3.14). Then it follows immediately that we have the

Proposition (4.7). For every point $\underline{t} \in \underline{H}_\mu$, we have

$$\tau(F_\mu(\underline{t})) = \mu\text{-rank } K(\underline{t}).$$

Put

$$\mu sm(F_\mu(\underline{t})) = \dim\{\tilde{\underline{t}}\in\underline{H}_\mu \mid \text{rank } K(\tilde{\underline{t}}) = \text{rank } K(\underline{t})\} - \text{rank } K(\underline{t})$$

$$\mu m(f) = \max_{\underline{t}\in\underline{H}_\mu} \mu sm(F_\mu(\underline{t})).$$

Notice that $\mu m(f)$ is the usual modality of f with respect to the μ-constant stratum, under the action of the contact groups, see [A]. If $\tau = \tau(F(\underline{t}))$ and

$$\underline{M}_{\mu\tau} = \text{im}\{\underline{H}_\mu \cap \underline{S}_\tau \to \underline{M}_\tau\}$$

then,

$$\mu sm(F_\mu(\underline{t})) = \dim \underline{M}_{\mu\tau,c}$$

where $\underline{M}_{\mu\tau,c}$ is the component of $\underline{M}_{\mu\tau}$ containing the class of $F_\mu(\underline{t})$. Put

$$\tau_{\min} = \min_{\underline{t}\in H_\mu} \tau(F_\mu(\underline{t}))$$

$$m_0(f) = \dim \underline{H}_\mu.$$

§5. PLANE CURVE SINGULARITIES WITH k^*-ACTION

Introduction. It turns out to be a big difference between the theory of hypersurface singularities in dimension 1 and in higher dimensions.

Even in the quasihomogenous case the situation in dimensions ≥ 2 seems to be horrendously complicated.

In dimension 1 however there is some light to be seen. In this § we study the dimensions of the components $\underline{M}_{\mu\tau}$, $\tau_{min} \leq \tau \leq \tau(f)$ of the μ-constant stratum of the local moduli suite for a quasihomogenous curve singularity f.

The main results are (5.2), where we prove that $\underline{M}_{\mu\tau} \neq \emptyset$ for all τ with $\tau_{min} \leq \tau \leq \mu(f) = \mu$, dim $\underline{M}_{\mu\tau} \geq$ dim $\underline{M}_{\mu,\tau+1}$ for all τ with $\tau_{min} \leq \tau \leq \tau(f)-1$, and (5.13), (5.14) and (5.16) where we give formulas for dim $\underline{M}_{\mu,\tau_{min}}$, generalizing results of Zariski [Z] and Delorme [Del].

We propose to prove the following theorem,

Theorem (5.1). Suppose $f \in k[x_1,x_2]$ is a weighted homogeneous polynomial defining an isolated singularity. Then
$\mu m(f) = m_0(f)+\tau_{min}-\mu$.

Obviously (5.1) is a consequence of the more precise

Theorem (5.2). Under the conditions of (5.1), let F_μ be a miniversal μ-constant family. Then

(i) $\underline{t} \to \mu sm(F_\mu(\underline{t}))$ is upper semicontinuous

(ii) $\tau(F_\mu(\underline{t}))$, $\underline{t} \in \underline{H}_\mu$ takes all values between τ_{min} and $\tau(f) = \mu$.

We shall restrict ourselves to the case $f = x_1^{a_1}+x_2^{a_2}$, $a_1 \leq a_2$. The other cases may be proved in the same manner.

Before we start the proof of (5.2), let us pause to explain the significance of these results. By (4.7) the subscheme $\underline{H}_\mu \cap \underline{S}_{\tau_{min}}$ of $\underline{H}_\mu$ is open and dense.

Talking about the generic μ-constant deformation of f, is therefore the same as talking about the τ_{min} substratum of $\underline{H}_\mu$.

Now, one would like to believe that this generic μ-constant deformation represents all but a strictly lower dimensional subset of the space of all isomorphism classes of μ-constant deformations of f.

This is, as we shall see, not true in general. What Theorem (5.1) tells us is that the generic μ-constant deformation of $f = x_1^{a_1}+x_2^{a_2}$ at least represents a subset of these isomorphism classes, of maximal dimension.

In dimension greater or equal to 2, the situation becomes much worse. In fact, for $f = x_1^3+x_2^{10}+x_3^{19}$ $\mu = 324$, $\tau_{min} = 246$, but $\tau = 247$, does not occur, and $\underline{S}_{248}$ is an open subset of a proper hypersurface of $\underline{H}_\mu$. Nevertheless, the dimension, dim $\underline{M}_{\mu,\tau_{min}}$ of the space of isomorphism classes corresponding to the generic μ-constant deformations, i.e. those with $\tau = 246$, is strictly less than the dimension of $\underline{M}_{\mu,248}$.

This shows that talking about the generic μ-constant deformation as being generic in the sense of moduli is, maybe o.k. in dimension 1, but certainly misleading in dimension ≥ 2.

Now, to prove (5.2) we shall have to study the matrix $K(\underline{t})$ more carefully. Notice first that (4.6) (ii), together with (3.11) implies $H_0 = H_\mu/J$, where J is the ideal generated by $\{t_{\underline{\alpha}} \mid |\underline{\alpha}|>1\}$. Thus $H_0 = k[t_{\underline{\alpha}}]$, $|\underline{\alpha}| = 1$.

Now let $\lambda_1<\ldots<\lambda_\mu$ be the monomial basis $\{\underline{x}^{\underline{\alpha}}\}_{\underline{\alpha}\in I}$ of $H^1 = k[[\underline{x}]]/(f,\frac{\partial f}{\partial x_1},\frac{\partial f}{\partial x_2})$ ordered by degree and lexicographic order, i.e. such that

$$\underline{x}^{\underline{\alpha}} < \underline{x}^{\underline{\beta}} \text{ or } \underline{\alpha} < \underline{\beta}$$

if either $|\underline{\alpha}|<|\underline{\beta}|$ or $|\underline{\alpha}| = |\underline{\beta}|$ and $\underline{\alpha} = (\underline{\alpha}_1,\alpha_2)$, $\underline{\beta} = (\underline{\beta}_1,\beta_2)$ with $\underline{\alpha}_1 < \underline{\beta}_1$. If $\lambda_i = \underline{x}^{\underline{\alpha}_i}$, we put $\deg \lambda_i = |\underline{\alpha}_i|$.

Remark (5.3). Observe that:

(i) $\lambda_1 = 1$, $\lambda_\mu = x_1^{a_1-2}\cdot x_2^{a_2-2}$

(ii) $E\lambda_i = 0$, if $\deg \lambda_i > \deg \lambda_\mu - \min\{\deg \lambda_k \mid \deg\lambda_k > 1\}$

(iii) duality; $\lambda_i \cdot \lambda_{\mu-i+1} = \lambda_\mu$. We shall write $\lambda_i^\vee = \lambda_{\mu-i+1}$.

(iv) $E\lambda_i$ is contained in the submodule of H^1, generated by $\{\lambda_k \mid \deg\lambda_k > 1\}$.

From this it follows that the matrix K looks like

$$\begin{array}{c} \\ r \\ \mu-r \end{array} \begin{pmatrix} r \\ K_0 \\ \hline 0 \end{pmatrix} \qquad \begin{aligned} r &= m_0(f) - sm(f) = \#\{k \mid \deg \lambda_k > 1\} \\ &= \tfrac{1}{2}(a_1-3)(a_2-3) + [\tfrac{a_2}{a_1}] - \tfrac{q+1}{2} + \begin{cases} 1 & a_1 = a_2 \\ 0 & \text{otherwise} \end{cases} \end{aligned}$$

with $q = (a_1, a_2)$

Let M_1 be the sub H_μ-module of $H_\mu[[x_1,x_2]]/(\frac{\partial F_\mu}{\partial x_1}, \frac{\partial F_\mu}{\partial x_2})$ generated by the λ_i's with $\deg \lambda_i \leq \deg \lambda_\mu - \min\{\deg \lambda_k \mid \deg \lambda_k > 1\}$, i.e. by the r first λ_i's, and let M_2 be the sub H_μ-module generated by the λ_i's with $\deg \lambda_i > 1$, i.e. by those $\underline{x}^{\underline{\alpha}}$'s sitting strictly above the face of f.

Notice that $\lambda_i \in M_1$ iff $\lambda_i^\vee \in M_2$. Put $\mu_j = \lambda_{r-j+1}^\vee$ and $t_j = t_{\underline{\alpha}_j}$; if $\mu_j = \underline{x}^{\underline{\alpha}_j}$. Then $\{\lambda_i\}_{i=1}^r$ is a basis for M_1, and $\{\mu_j\}_{j=1}^r$ is a basis for M_2.

Observe also that:

(v) for all i, $1 \leq i \leq r$ there is a j_i with the following property.

(1) if $j \geq j_i$ then there is a $k(i,j)$, $1 \leq k(i,j) \leq r$ such that $\mu_j = \mu_{k(i,j)} \cdot \lambda_i$

(2) if $\mu_j = \mu_k \cdot \lambda_i$ then $j \geq j_i$.

With these notations we find that $K_0 = (h_{ij})$ is the matrix associated to the H_μ-linear map

$$E : M_1 \to M_2$$

defined by multiplication with E. Before we proceed with the general theory, let us consider an example.

<u>Example (5.4)</u>. Let $f = x^5+y^{11}$. Then

$$F_\mu(\underline{t}) = f+t_1xy^9+t_2x^2y^7+t_3x^3y^5+t_4x^2y^8+t_5x^3y^6$$

$$+ t_6x^2y^9+t_7x^3y^7+t_8x^3y^8+t_9x^3y^9$$

$$E = \frac{1}{55}(t_1xy^9+2t_2x^2y^7+3t_3x^3y^5+7t_4x^2y^8+8t_5x^3y^6+12t_6x^2y^9$$

$$+ 13t_7x^3y^7+18t_8x^3y^8+23t_9x^3y^9)$$

This is easily seen by inspecting the Newton diagram of f.

$a_1 = 5 \quad a_2 = 11$

$\mu(f) = \tau(f) = 40$

$m_0(f) = 9$

$sm(f) = 0$

$r = 9,\ H_\mu = k[t_1,\ldots,t_9]$

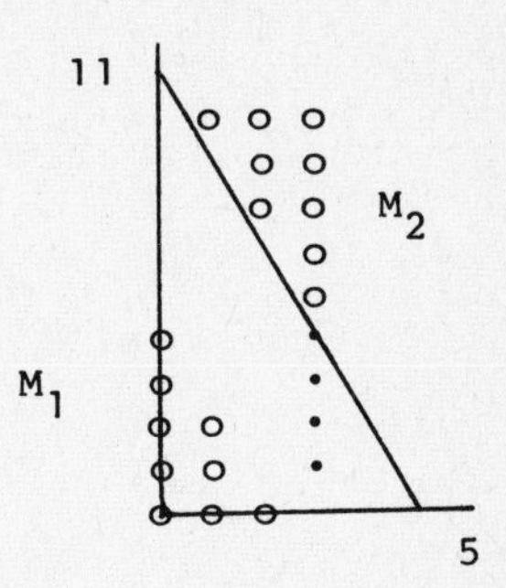

Now, put

$$A = 2t_2 - \frac{9}{11}t_1^2,$$

$$B = 3t_3 - \frac{7}{11}t_1t_2,$$

$$C = 7t_4 + \frac{1}{11}t_1t_3\left(\left(\frac{9}{11}\right)^2t_1^3 - \frac{25}{11}t_1t_2+3t_3\right),$$

$$D = 8t_5 - \frac{8}{11}t_1t_4 + \frac{1}{11}t_1t_3\left(\frac{63}{11^2}t_1^2t_2 - \frac{5}{11}t_1t_3 - \frac{14}{11}t_2^2\right)$$

$$E \equiv 8t_5 - \frac{71}{11}t_1t_4 - \frac{5}{11^2}t_1^2t_3^2,\quad F \equiv 7t_4 + \frac{3}{11}t_1t_3^2 \qquad \text{mod } (A,B)$$

then the matrix $55 \cdot K_0$ looks like:

t_1	$2t_2$	$3t_3$	$7t_4$	$8t_5$	$12t_6$	$13t_7$	$18t_8$	$23t_9$
0	0	0	A	B	C	D	*	*
0	0	0	0	0	A	B	E	*
0	0	0	0	0	t_1	$2t_2$	F	*
0	0	0	0	0	0	0	$B-\frac{9}{11}t_1A$	*
0	0	0	0	0	0	0	A	*
0	0	0	0	0	0	0	0	$B-\frac{9}{11}t_1A$
0	0	0	0	0	0	0	0	A
0	0	0	0	0	0	0	0	t_1

The flattening stratification $\{\underline{S}_\tau\}_{\tau\in T_\mu}$ of $A^1(H_\mu,\tilde{A}_\mu;\tilde{A}_\mu)$, see § 3, coincides with the rank-filtration of K_0, and it is easily seen that $T_\mu = \{34,35,36,37,38,39,40\}$ is the set of possible Tjurina numbers with constant μ $(= 40)$ in the neighbourhood of f.

As T. Yano observed, a suitable change of the monomial base of M_1 leads to a symmetric matrix. In our example we find that $55\cdot K_0$ becomes,

$$\begin{array}{ccccccccc}
t_1 & 2t_2 & 3t_3 & 7t_4 & 8t_5 & 126_6 & 13t_7 & 18t_8 & 23t_9 \\
 & & & A & B & C & D & E & 18t_8 \\
 & & & & & 2t_2 & 3t_3+t_1t_2 & D & 13t_7 \\
 & & & & & t_1 & 2t_2 & C & 12t_6 \\
 & & & & & & & B & 8t_5 \\
 & & & & & & & A & 7t_4 \\
 & & & & & & & & 3t_3 \\
 & & & & & & & & 2t_2 \\
 & & & & & & & & t_1
\end{array}$$

with $A = 2t_2 - \frac{9}{11}t_1^2 \quad B = 3t_3 - \frac{7}{11}t_1t_2 \quad C = 7t_4 + \frac{3}{11}t_1t_3^2$

$$D = 8t_5 - \frac{8}{11}t_1t_4 + \frac{2}{11}t_1^2t_3^2$$

$$E = 13t_7 - \frac{117}{11}t_1t_6 + \frac{3}{11}t_1^2t_3t_5 + \frac{55}{11^2}t_1t_2t_3t_4 + \frac{7}{5\cdot 11^3}t_1^3t_2^2t_3^2$$

The reduced flattening stratification is given by:

$\underline{S}_{34}$: $4t_2^2-3t_1t_3-t_1^2t_2 \neq 0$

$\underline{S}_{35}$: $4t_2^2-3t_1t_3-t_1^2t_2 = 0$ and $A\neq 0$ or $B\neq 0$ or

$D(2t_2C-t_1D)-C(C(3t_3+t_1t_2)-2t_2D) \neq 0$.

Notice that the last minor defining $\underline{S}_{35}$ is not invariant, but nevertheless any point in $\underline{S}_{35}$ has an open invariant affine neighbourhood.

In fact, replacing the last minor by $(t_1D-2t_2C)^2-2(A^2+\frac{1}{11}t_1^2A)(t_1t_7-2t_2t_6)$ we get an open invariant covering. $\underline{S}_{35}$ is smooth.

$\underline{S}_{36}$: $A = B = t_1(9t_1C-11D) = 0$ and C^2-t_1E or $D^2-(\frac{9}{11})^2 t_1^3E \neq 0$.

$\underline{S}_{37}$: $A = B = C^2-t_1E = D- \frac{9}{11}t_1C = 0$ and $t_1 \neq 0$ or $E \neq 0$.

$\underline{S}_{38}$: $t_1 = \ldots = t_5 = t_7 = 0$ and $t_6 \neq 0$ or $t_8 \neq 0$.

$\underline{S}_{39}$: $t_1 = \ldots = t_8 = 0$ and $t_9 \neq 0$.

$\underline{S}_{40}$: $t_1 = \ldots = t_9 = 0$.

Remark. Notice that $\underline{S}_{\tau_{min}}$ is not necessarily the τ_{min}-constant stratum of $\underline{H}$: In fact the family $x_1^5+x_2^{11}+tx_1^2x_2^7+2t_1x_1^4x_2^2+t_1x_1^2x_2^{34}$ is τ-const. ($\tau = 34$ for $0<|t|<<1$) but not μ-const. ($\mu(t_1 \neq 0) = 39$).

We find the table

τ	34	35	36	37	38	39	40
$\tau m(f)$	3	3	2	2	1	0	0

Thus, in particular,

$$\mu m(f) = 3.$$

The nice properties of the matrix K_0 becomes apparent if one restricts attention to the linear terms of $K_0(\underline{t})$. The corresponding matrix will be denoted $L(\underline{t})$. In fact $55 \cdot L(\underline{t})$ looks like:

$$\begin{matrix}
t_1 & 2t_2 & 3t_3 & 7t_4 & 8t_5 & 12t_6 & 13t_7 & 18t_8 & 23t_9 \\
0 & 0 & 0 & 2t_2 & 3t_3 & 7t_4 & 8t_5 & 13t_7 & 18t_8 \\
0 & 0 & 0 & 0 & 0 & 2t_2 & 3t_3 & 8t_5 & 13t_7 \\
0 & 0 & 0 & 0 & 0 & t_1 & 2t_2 & 7t_4 & 12t_6 \\
0 & 0 & 0 & 0 & 0 & 0 & 0 & 3t_3 & 8t_5 \\
0 & 0 & 0 & 0 & 0 & 0 & 0 & 2t_2 & 7t_4 \\
0 & 0 & 0 & 0 & 0 & 0 & 0 & 0 & 3t_3 \\
0 & 0 & 0 & 0 & 0 & 0 & 0 & 0 & 2t_2 \\
0 & 0 & 0 & 0 & 0 & 0 & 0 & 0 & t_1
\end{matrix}$$

Notice that $L(\underline{t})$ is symmetric on the antidiagonal with respect to the monomial base we started with.

Now this symmetry is a general feature. In fact let $L(\underline{t}) = (\ell_{ij})$, then we have the following

Proposition (5.5). With the notations above,

(i) $\ell_{ij} = \ell_{r-j+1,r-i+1}$

(ii) $\ell_{ij} = d_{k(i,j)} t_{k(i,j)}$ with $d_i = \deg \mu_i - 1$, whenever $j > j_i = \min\{j, h_{ij} \neq 0\}$

(iii) $k(i,j) < k(i,j+1)$ and $k(i+1,j) < k(i,j)$

(iv) $j_i < j_{i+1}$.

Proof. (i) Since $E\lambda_i = \sum_{j=1}^{r} h_{ij}\mu_j$ and since $\lambda_{r-\ell+1} = \check{\mu}_\ell$ we find

$$E\lambda_i\lambda_{r-\ell+1} = \sum_{j<\ell} h_{ij}\mu_j\check{\mu}_\ell + h_{i\ell}\mu_r.$$

Since $E\lambda_i\lambda_{r-\ell} = E\lambda_{r-\ell}\lambda_i$ and since the linear part of $\sum_{j<\ell} h_{ij}\mu_j\check{\mu}_\ell$ is of the form $\sum_{s<r} *\mu_s$ it follows that $\ell_{ij} = \ell_{r-j+1,r-i+1}$, i.e. (i). Consider (4.6), (ii), and (5.3) (v), then the rest follows.

Q.E.D.

Proposition (5.6). There is a basis $\{\bar{\lambda}_i\}_{1<i<r}$ of M_1 with the following properties:

(1) $\bar{\lambda}_i = \lambda_i + \sum_{j>i} e_{ij}\lambda_j$, e_{ij} homogeneous of degree $\deg\lambda_i - \deg\lambda_j$.

(2) The matrix of $E: M_1 \to M_2$ with respect to the bases $\{\bar{\lambda}_i\}$ in M_1 and $\{\mu_i\}$ in M_2 is symmetric with linear part $L(\underline{t})$.

Proof: Consider on $H_\mu[[x_1,x_2]]/(\frac{\partial F_\mu}{\partial x_1}, \frac{\partial F_\mu}{\partial x_2})$ the following pairing defined by the coefficient of the Hessian λ_μ of the product of two elements: Let $h \in H_\mu[[x_1,x_2]]/(\frac{\partial F_\mu}{\partial x_1}, \frac{\partial F_\mu}{\partial x_2})$, define $c_\mu(h)$ by $h = \sum c_i(h)\lambda_i$, and put

$\langle h,k\rangle := c_\mu(h\cdot k)$

$\langle,\rangle$ is a nondegenerate symmetric bilinear form:

(i) $\langle\lambda_i,\lambda_{\mu-i+1}\rangle = 1$

(ii) $\langle\lambda_i,\lambda_j\rangle = 0$ if $i+j > \mu+1$

(iii) if $\langle\lambda_i,\lambda_j\rangle \neq 0$ then $\langle\lambda_i,\lambda_j\rangle$ is homogeneous of degree $\deg\lambda_i + \deg\lambda_j - \deg\lambda_\mu$.

Denote by E_0 the matrix of $\langle,\rangle$ with respect to the base $\{\lambda_i\}$. Now obviously the map defined by multiplication with F_μ is selfadjoint, i.e. $\langle\lambda_i F_\mu,\lambda_j\rangle = \langle\lambda_i,\lambda_j F_\mu\rangle$. Denote by C the matrix of $\langle F_\mu -,-\rangle$ with respect to the base $\{\lambda_i\}$. Notice that $\langle\lambda_i F_\mu,\lambda_j\rangle = 0$ if $\lambda_i\cdot\lambda_j \notin \{\lambda_1,\dots\dots,\lambda_r\}$ (this is a special property of the curve case). Now

(iv) $C = K\cdot E_0$.

This implies that $E_0^{-1}\,C\,E_0^{-1} = E_0^{-1}K$ is symmetric. E_0^{-1} defines a base change in $H_\mu[[x_1,x_2]]/(\frac{\partial F_\mu}{\partial x_1},\frac{\partial F_\mu}{\partial x_2})$. The induced base change in M_1 has the required properties:

Let $E_0^{-1} = (\bar e_{ij})$ then

$\bar e_{i,\mu-i+1} = 1$

$\bar e_{i,j} = 0,\ i+j < \mu+1$

$\deg \bar e_{ij} = \deg\check\lambda_i + \deg\check\lambda_j - \deg\lambda_\mu$.

Define $\{\bar\lambda_i\}$ by

$\bar\lambda_i = \sum_{j<r} \bar e_{j,\mu-i+1}\lambda_j$.

Let $(\bar c_{ij})$ be the matrix of $E_0^{-1}CE_0^{-1}$ then, because $\lambda_j E = 0$ if $j>r+1$, the matrix of $E: M_1 \to M_2$ with respect to $\{\bar\lambda_i\}$ in M_1 and $\{\mu_j\}$ in M_2 defined by $\bar\lambda_i E = \sum_{j=1}^{r} \bar h_{ij}\mu_j$ satisfies $\bar h_{ij} = \bar c_{i,r-j+1}$.

Q.E.D.

Notice that a base change in M_1 gives new generators of the kernel of the Kodaira-Spencer map as the corresponding combination of the $\delta_{\underline{\alpha}_i}$.

Corollary (5.7). There is a basis $\{\bar{\mu}_i\}_{1<i<r}$ of M_2 and an automorphism $\phi : H_\mu \to H_\mu$ with the following properties:

(1) $\bar{\mu}_i = \mu_i + \sum_{j>i} e_{ij}\mu_j$, $\deg e_{ij} + \deg \mu_j = \deg \mu_i$

(2) ϕ is homogeneous

(3) the matrix of $E : M_1 \to M_2$ with respect to the basis $\{\lambda_i\}$ in M_1 and $\{\bar{\mu}_i\}$ in M_2 is $L(\phi(\underline{t}))$.

Proof: Consider the base change on M_2 induced by E_0:

$$\bar{\lambda}_{\mu-r+i} = \bar{\mu}_i = \sum_j \bar{e}_{j,r-i+1}\lambda_j = \sum_{j>i} \bar{e}_{\mu-r+j,r-i+1}\mu_j$$

because of

$$\bar{e}_{j,r-i+1} = 0 \quad \text{if} \quad r-i+1+j < \mu+1 \quad \text{and} \quad \lambda_{\mu-r+j} = \mu_j .$$

The matrix of $E : M_1 \to M_2$ defined by $\lambda_i E = \sum_{j=1}^{r} \bar{h}_{ij}\bar{\mu}_j$ satisfies $\bar{h}_{ij} = c_{i,r-j+1}$. Now consider $\lambda_i E = \sum_{j=1}^{r} h_{ij}\mu_j$ then

$$c_{ij} = \langle \lambda_i F_\mu, \lambda_j \rangle = \begin{cases} h_{\ell,r} & \text{if} \quad \lambda_i\lambda_j = \lambda_\ell, \ \ell<r \\ 0 & \text{otherwise} \end{cases} .$$

Notice that with the notations of (5.5) $\ell = k(i,r-j+1)$.
But $h_{\ell,r} = (\deg \mu_{r-\ell+1} - 1)t_{r-\ell+1} + \tilde{h}_\ell(t_1,\ldots,t_{r-\ell})$. We may choose $\phi : H_\mu \to H_\mu$ with $\phi((\deg \mu_{r-\ell+1} - 1)t_{r-\ell+1}) = h_{\ell,r}$.

Q.E.D.

Remark: If we consider the corresponding situation for surfaces then $c_{ij} = \langle \lambda_i F_\mu, \lambda_j \rangle$ may be different from zero when $\lambda_i\lambda_j$ is not in the monomial base. This is the reason why the flattening stratification cannot be described in terms of the linear matrix and why (5.2) fails in higher dimensions. In the example $x^3 + y^{10} + z^{19}$ the minor of the matrix K_0 giving maximal rank (= 78) is

$$
\begin{array}{ccccccccc}
-\frac{1}{3}t_1^2 & -\frac{5}{3}t_1t_2 & * & * & * & * & * & * & * \\
0 & -\frac{1}{3}t_1^2 & * & & & & & & * \\
\cdot & 0 & * & & & & & & * \\
\cdot & & & * & & & & & * \\
\cdot & & & & * & & & & * \\
 & & & & & & & & * \\
\cdot & & & & & & 0 & -\frac{1}{3}t_1^2 & -\frac{5}{3}t_1t_2 \\
0 & 0 & \cdot & & & & \cdot & 0 & -\frac{1}{3}t_1^2
\end{array}
$$

For $t_1 = 0$ the rank decreases by 2.

Proof of (5.2). We first prove the second part. Because of (5.7) the rank-filtration of $K_0(\underline{t})$ is isomorphic to that of $L(\underline{t})$. So let's compute the minors of $L(\underline{t})$.

We shall use (5.5). In particular, it follows that any minor of the linear matrix $L(\underline{t})$ has the form,

$$M(\underline{t}) = (d_{a(i,j)}t_{a(i,j)})$$

where $a(i,j) \in \{0,1,\ldots,r\}$, $i,j = 1,\ldots,m$, $d_{a(i,j)} = \deg \lambda_{a(i,j)}^{-1} \neq 0$, if $a(i,j) \neq 0$, $d_0 = 0$, and where

(1) $a(i,j-1) \neq 0$ implies $a(i,j) < a(i,j+1)$

(2) $a(i-1,j) \neq 0$ implies $a(i,j) < a(i-1,j)$.

Lemma (5.8). With the notations above,

$$\det M(\underline{t}) \neq 0 \text{ if and only if } a(i,i) \neq 0 \text{ for all } i = 1,\ldots,m.$$

Proof. Suppose $a(i,i) = 0$ for some i, then $M(\underline{t})$ has the form

$$
\begin{array}{c}
 \\ \\ \\ \\ i \; - \\ \\ \\
\end{array}
\begin{pmatrix}
 & & i & & \\
 & & \vdots & & \\
 & & * & & \\
 & & \vdots & & \\
 & & * & & \\
0 & \cdots & 0 & * \cdots & * \\
\vdots & & 0 & & \\
\vdots & & \vdots & & \\
0 & \cdots & 0 & &
\end{pmatrix}
$$

therefore $\det M(\underline{t}) = 0$.

Assume $a(i,i) \neq 0$ for all $i = 1,\ldots,m$. Use induction on the number of $a(i,j) \neq 0$. Let $s = \min\{a(i,j) \mid a(i,j) \neq 0\}$, and assume

$$\det M(\underline{t})_{\{t_s=0\}} = 0.$$

By induction $M(\underline{t})_{\{t_s=0\}}$ has a diagonal element $\bar{a}(i,i) = 0$. This implies that $M(\underline{t})$ has the form

$$\begin{array}{c} \\ \\ \\ \\ i\ - \\ \\ \\ \\ \end{array}\left(\begin{array}{cccccc} & & & & i & \\ & & & & \vdots & \\ & & & & * & \\ & A & & & \vdots & \\ & & & & * & \\ 0 & \cdots & 0 & d_s & \cdot t_s & * \cdots * \\ 0 & \cdots & \cdot & & 0 & \\ \vdots & & \vdots & & \vdots & B \\ 0 & \cdots & 0 & & 0 & \end{array}\right)$$

where A and B are minors of $L(\underline{t})$, and therefore has the same form as $M(\underline{t})$. By induction $\det A \neq 0$, $\det B \neq 0$, therefore

$$\det M(\underline{t}) = d_s \cdot t_s \det A \cdot \det B \neq 0. \qquad \text{Q.E.D}$$

If we agree to call "a diagonal" any string of elements of a matrix parallel to <u>the diagonal</u>, we may prove the following.

<u>Lemma (5.9)</u>. The rank of $L(\underline{t})$ is the length ℓ of the maximal diagonal containing no zeros. With the notations of (5.5):

$$\ell = m_0(f) - sm(f) - \max_i \{j_i - i\}.$$

<u>Proof</u>. From (5.5) and (5.8) we deduce $\ell \leq \operatorname{rank} L(\underline{t})$. To prove the inverse inequality, let $M(\underline{t})$ be any $m\times m$ minor of $L(\underline{t})$ gotten by picking the i_1'th, i_2'nd, ..., i_m'th rows and the j_1'th, j_2'nd, ..., j_m'th. column of $L(\underline{t})$. Using (5.5) and (5.8) we find that $\det M(\underline{t}) \neq 0$ implies that $\ell_{i,r} \neq 0$ for $1 \leq i \leq i_m$, $\ell_{i,r-1} \neq 0$ for $1 \leq i \leq i_{m-1}, \ldots, \ell_{i,r-m} \neq 0$ for $1 \leq i \leq i_1$. Therefore there exists a diagonal of $L(\underline{t})$ of length m containing no zeros.

Q.E.D.

The second part of (5.2) now follows by induction on τ. Let's consider the maximal diagonal of $L(\underline{t})$ containing no zeros. Among the $k(i,j)$ for which $t_{k(i,j)}$ occur on the corresponding diagonal of $L(\underline{t})$, let $k(i_\ell, j_\ell)$ be the smallest. On the subset defined by $t_i = 0$, $i \leq k(i_\ell, j_\ell)$ the rank of $L(\underline{t})$ has decreased by 1. To prove (5.2) (i), we need the following lemma.

Lemma (5.10). Let $0 \leq p \leq \max_{\underline{t}}\{\text{rk } L(\underline{t})\} = p_0$, then

$$\dim\{\underline{t} \in \underline{H} \mid \text{rk } L(\underline{t}) \leq p_0 - p\} \leq r-p.$$

Proof. Consider the ideal I_p of H_μ generated by the (p+1)-minors of $L(\underline{t})$. We need only prove:

Lemma (5.11). Let $a(i,j) \in \{0,\ldots,r\}$, $i,j = 1,\ldots,m$ satisfy

(1) $a(i,j+1) \neq 0$ implies $a(i,j) < a(i,j+1)$

(2) $a(i-1,j) \neq 0$ implies $a(i,j) < a(i-1,j)$

and let $d_0 = 0$, $d_s = \deg \mu_s - 1$ as above. Consider $M(\underline{t}) = (d_{a(i,j)} t_{a(i,j)})$, and assume $\det M(\underline{t}) \neq 0$. Let I_p be the ideal generated by the (p+1)-minors of $M(\underline{t})$. Then

$$\text{ht } I_p \geq m-p \quad \text{i.e.} \quad \dim k[\underline{t}]/I_p \leq r-m+p.$$

Proof. Use induction on m and on the number of different $t_{a(i,j)}$'s involved in the matrix.

Let $s = \min\{a(i,j) \mid a(i,j) \neq 0\}$ and let U be a component of $V(I_p) = \{\underline{t} \in \underline{H} \mid \forall\, \underline{p} \in I_p, \underline{p}(\underline{t}) = 0\}$.

1. case. $U \subseteq V(t_s) = \{\underline{t} \mid t_s = 0\}$. Consider the sub-matrix $(d_{a(i,j)} t_{a(i,j)})_{1 \leq i \leq m-1,\, 2 \leq j \leq m}$ obtained from $M(\underline{t})$ by deleting the mth row and the 1st column. The conditions (1) and (2) together with the assumption $\det(d_{a(i,j)} \cdot t_{a(i,j)}) \neq 0$ imply that $s \neq a(i,i+1)$ $i = 1,\ldots,m-1$. Therefore $\det(d_{a(i,j)} t_{a(i,j)})_{1 \leq i \leq m-1,\, 2 \leq j \leq m} \neq 0$, and we may apply the induction hypotheses.

2. case. $U \not\subseteq V(t_s)$. It follows from (1) and (2) that $\operatorname{rk} M(\underline{t}) \geq \ell$ for all $\underline{t}$ with $t_s \neq 0$, where ℓ is the number of times t_s occurs in $M(\underline{t})$, see fig. In particular this implies $\ell \leq p$. Consider the $(m-\ell)\times(m-\ell)$ sub-matrix $M_0(\underline{t})$ of $M(\underline{t})$ obtained by deleting the rows and columns in which t_s occurs.

$$M(\underline{t}) = \left(\begin{array}{ccccccccccccccc}
* &&&&&&&&&&&&&& \\
t_s & * &&&&&&&&&&&&& \\
0 & * & * & * &&&&&&&&&&& \\
0 & 0 & 0 & t_s & * &&&&&&&&&& \\
 && 0 & 0 & * &&&&&&&&&& \\
 &&& 0 & * & * &&&&&&&&& \\
 &&& 0 & 0 & t_s & * &&&&&&&& \\
 &&&&& 0 & * &&&&&&&& \\
 &&&&& 0 & * &&&&&&&& \\
 &&&&& 0 & * & * & * & * & * & * &&& \\
 &&&&& 0 & 0 & 0 & 0 & 0 & 0 & t_s & * && \\
 &&&&&&&&&&& 0 & * && \\
 &&&&&&&&&&& 0 & * && \\
 &&&&&&&&&&& 0 & * & * & \\
 &&&&&&&&&&& 0 & 0 & t_s & *
\end{array}\right)$$

For $t_s \neq 0$ it is easy to see that $\operatorname{rk} M(\underline{t}) = \ell + \operatorname{rk} M_0(\underline{t})$, therefore

$$\{\underline{t} \mid \operatorname{rk} M(t) < p\} = \{\underline{t} \mid \operatorname{rk} M_0(\underline{t}) < p-\ell\}.$$

Note that since $\det M(\underline{t}) \neq 0$, all t_s occuring in $M(\underline{t})$ sit under the diagonal, thus $\det M_0(\underline{t}) \neq 0$. Moreover $M_0(\underline{t})$ does not contain t_s. We may therefore apply the induction hypotheses, and the lemma is proved. Q.E.D.

This ends the proof of theorem (5.2).

To compute the maximal rank of $K_0(\underline{t})$, or what is the same, the dimension of $M_{\tau_{min}}$, i.e. the number $\mu m(f)$ defined above, it turns out that one may use (5.2). In fact the maximal rank of $K_0(\underline{t})$ is the same as the maximal rank of $L(\underline{t})$ which is the same as the dimension of the maximal orbits of the action ρ of $\operatorname{Der}(k[[\underline{x}]]/(f))/\operatorname{Der}_\mu$ on the tangent space of $\underline{H}_\mu$ at $\underline{0}$, which is a subspace of $A^1(k,k[[\underline{x}]]/(f) \to k;k[[\underline{x}]]/(f))$. In the case of $f = x_1^{a_1} + x_2^{a_2}$, Hans Olav Herøy has proved the following formula, see [Her].

Proposition (5.12). (Herøy). Let $f = x_1^{a_1} + x_2^{a_2}$ with $2|(a_1,a_2)$ and $a_1 \leq a_2$ and let $q = (a_1,a_2)$. Then the maximal dimension of the orbits of ρ is

$$\mathrm{rk}\, K_0 = \frac{(a_1-2)(a_2-2)}{4} - \frac{q}{2} + \begin{cases} 1 & \text{if } q = a_1 \\ 0 & \text{otherwise} \end{cases}$$

Corollary (5.13). With the above notations, assuming $2|(a_1,a_2)$ we have

$$\mu m(f) = m_0(f) - \frac{(a_1-2)(a_2-2)}{4} + \frac{q}{2} - \begin{cases} 1 & \text{if } q = a_1 \\ 0 & \text{otherwise} \end{cases}$$

$$= sm(f) + \frac{1}{4}(a_1-4)(a_2-4) - 1 + \left[\frac{a_2}{a_1}\right] - \begin{cases} 1 & \text{if } q = a_1,\ a_1 < a_2 \\ 0 & \text{otherwise} \end{cases}$$

When $(a_1,a_2) = 1$, Delorme [Del] gives a formula for rank K_0:

Let $\frac{a_2}{a_1} = [r_1,\ldots,r_k] = r_1 + \cfrac{1}{r_2 + \cfrac{1}{r_3 + \cdots + \cfrac{1}{r_k}}}$

define ℓ_i and t_i inductively:

$\ell_k = 0$, $t_k = 1$

$\ell_{i-1} = \ell_i + t_i r_i$ and $t_{i-1} = \begin{cases} 0 & \text{if } t_i = 1 \text{ and } \ell_{i-1} \text{ even} \\ 1 & \text{otherwise} \end{cases}$

Proposition (5.14): (1) The maximal dimension of the orbits of ρ is

$$\mathrm{rk}\, K_0 = \frac{(a_1-2)(a_2-2)}{4} - \frac{1}{4}\ell_0 + \frac{1}{2}t_1(r_1+t_2-2) + \frac{1}{2}$$

(2) $\mu m(f) = \frac{1}{4}(a_1-4)(a_2-4) + \frac{1}{4}\ell_0 + \frac{1}{2}(2-t_1)(r_1-2) - \frac{1}{2}t_1 t_2$

(3) $\frac{1}{4}(a_1-4)(a_2-4) \leq \mu m(f) \leq \frac{1}{4}(a_1-2)(a_2-4)$

except in the trivial case $a_1 = 2$, $a_2 = 3$.

Remark: The left hand side of the inequality is sharp: (8,11). The right hand side can be replaced by $\frac{1}{4}(a_1-3)(a_2-3)$.

Corollary (5.15):

(1) If $a_2 = ra_1+1$ or $a_2 = ra_1-1$

then the maximal dimension of the orbits is

$\frac{1}{4}(a_1-2)(a_2-3)$ a_1 even

$\frac{1}{4}(a_1-1)(a_2-r-3)$ a_1 odd

(2) If $a_2 = ra_1+2$
the maximal dimension of the orbit is

$\frac{1}{4}(a_1-2)(a_2-2)-1$ a_1 even

$\frac{1}{4}(a_1-1)(a_2-r-2)-1$ a_1 odd

(3) If $a_2 = ra_1-2$
the maximal dimension of the orbits is

$\frac{1}{4}(a_1-2)(a_2-2)-1$ a_1 even

$\frac{1}{4}(a_1-1)(a_2-r-2)$ a_1 odd

(4) If $a_2 = ra_1$
the maximal dimension of the orbits is

$\frac{1}{4}(a_1-2)(a_2-4)$ a_1 even

$\frac{1}{4}(a_1-1)(a_2-r-4)$ a_1 odd, $r \geq 2$

$\frac{1}{4}(a_1-3)^2$ a_1 odd, $r = 1$

Remark (5.16). (5.9) gives an algorithm to compute $\mu m(f)$ in the cases not covered by the above formulas. In fact, to find

$$\mu m(f) = \max_i \{j_i - i\}:$$

(1) Compute the spectral numbers corresponding to $\{\lambda_i\}$, $\{\mu_i\}$ (i.e. the weighted degrees): $s(\lambda_1),\ldots,s(\lambda_r)$, $s(\mu_1),\ldots,s(\mu_r)$. Let $\Lambda = \{s(\lambda_1),\ldots,s(\lambda_r)\}$, $M = \{s(\mu_1),\ldots,s(\mu_r)\}$.

(2) Then $j_i = \min\{j, s(\mu_j)-s(\lambda_i) \in M\}$.

For the example $x_1^5+x_2^{11}$

$\Lambda = \{0,5,10,11,15,16,20,21,22\}$

$M = \{56,57,58,62,63,67,68,73,78\}$

$j_1 = 1,\ j_2 = 4,\ j_3 = 6,\ j_4 = 6,\ j_5 = 8,\ j_6 = 8,$

$j_7 = j_8 = j_9 = 9.$

Remark (5.17). In the general case, Briançon, Granger and Maisonobe, have just found a recursive formula, reducing the computation of $\mu m(f)$ to the situation of (5.12) or (5.14), (see [Br]).

§6 THE GENERIC COMPONENT OF THE LOCAL MODULI SUITE

<u>Introduction</u>. In this § we shall return to the problem of constructing a local moduli suite for an isolated curve singularity f. As we have alrady pointed out, this problem has been studied by O. Zariski [Z] for $f = x_1^m + x_2^{m+1}$, and before him by S. Ebey [Eb].

We shall now prove that there exists a coarse moduli scheme for irreducible plane curve singularities with fixed semigroup $\Gamma = \langle a_1, a_2 \rangle$, $(a_1, a_2) = 1$, (see [Z] chap. II for definitions), and minimal τ. Notice that the assumption $(a_1, a_2) = 1$ is not necessary - see [L-M-P].

As we shall see, the existence of this moduli scheme follows from proving that the μ-constant stratum of the generic component $\underline{M}_{\tau_{min}}$ of the local moduli suite of $f = x_1^{a_1} + x_2^{a_2}$ is a scheme.

Merle [Me], and Washburn [Wash], have already published results of this type. The paper of Merle contains an error, and his main theorem (3.3) is false, see the counterexample §7, $\tau = 35$. Washburn's proof is incomplete. More seriously, his approach does not function. In particular his dimension formula is incorrect.

Washburn's method is based on the assumption that the multiplication $E: M_1 \to M_2$ (see §5) is compatible with the filtrations induced by the (x_1, x_2)-adic filtration of $k[[x_1, x_2]]$. This is unfortunately false in general, as one may see considering the example $f = x_1^5 + x_2^{12}$. We shall nevertheless show that there exists a filtration of this type, depending on (a_1, a_2). Using this filtration we show that there are enough invariants in $S_{\tau_{min}}$ under the action of $V_{\tau_{min}}$, to establish the existence of a geometric quotient $\underline{S}_{\tau_{min}} / V_{\tau_{min}}$, see [D-R] for definitions. We shall assume $k = C$, the field of complex numbers.

The purpose of this § is the proof of the following

<u>Theorem (6.1)</u>. Let $\Gamma = \langle a_1, a_2 \rangle$, $(a_1, a_2) = 1$, be a semigroup. There exists a coarse moduli space $\underline{T}_{\Gamma,\tau}$ parametrising all plane curve singularities with the semigroup Γ and minimal

Tjurina number $\tau = \tau_{min}$, and a universal family $\pi : X_{\Gamma,\tau} \to \underline{T}_{\Gamma,\tau}$ such that

(1) $\underline{T}_{\Gamma,\tau}$ is a quasismooth scheme, i.e. locally an open subset in a weighted projective space of dimension

$\frac{1}{4}(a_1-4)(a_2-4) + \frac{1}{4}\ell_0 + \frac{1}{2}(2-t_1)(r_1-2) - \frac{1}{2}t_1t_2$

(see (5.14) for the definition of ℓ_0, r_1, t_1, t_2).

(2) $X_{\Gamma,\tau}$ is an algebraic space and there is an affine covering $\{U_i\}$ of $\underline{T}_{\Gamma,\tau}$ such that $\pi^{-1}(U_i)$ are affine schemes.

Proof: We may suppose $a_1 < a_2$. Let $(X_0, 0)$ be a germ of an irreducible plane curve with singularity at 0, having Γ as semigroup. Consider $\tilde{X}_\mu \to \underline{H}_\mu$, the versal μ-constant deformation of the singularity defined by $f = x_1^{a_1} + x_2^{a_2}$. There is a $\underline{t} \in \underline{H}_\mu$ such that $(X_0, 0) \simeq (X_\mu(\underline{t}), 0)$ (cf. [A]).

Lemma (6.2). If for $\underline{t}_1, \underline{t}_2 \in \underline{H}_\mu$ $(X_\mu(\underline{t}_1), 0) \simeq (X_\mu(\underline{t}_2), 0)$ then $\underline{t}_1$ and $\underline{t}_2$ are in an analytically trivial subfamily of $X_\mu \to \underline{H}_\mu$.

Proof: The C^*-action induces a canonical filtration on the automorphism group Σ of $C[[x_1, x_2]]$. Put,

$\Sigma_\ell = \{\phi \in \Sigma \mid \deg(\phi(x_1) - x_1) \geq \ell + \frac{1}{a_1}, \deg(\phi(x_2) - x_2) \geq \ell + \frac{1}{a_2}\}$ and $\deg \phi = \ell$ iff $\phi \in \Sigma_\ell \setminus \Sigma_{\ell + \frac{1}{a_1a_2}}$. Suppose now that $(X_\mu(\underline{t}_1), 0) \simeq (X_\mu(\underline{t}_2), 0)$, i.e. suppose there is an $\phi \in \Sigma$ and a unit $u(x) \in C[[x_1, x_2]]$ such that

$$F_\mu(x_1, x_2, \underline{t}_1) = u(x) F_\mu(\phi(x_1), \phi(x_2), \underline{t}_2).$$

We shall prove that $\deg \phi \geq 0$.

If this is true $\underline{t}_1$ and $\underline{t}_2$ are in an analytically trivial subfamily of $\tilde{X}_\mu \to \underline{H}_\mu$. In fact using the C^*-action we may assume $\deg \phi > 0$. The trivial family of singularities

$$G(\lambda) := u(\lambda^{a_2}x_1, \lambda^{a_1}x_2) F_\mu\left(\frac{1}{\lambda^{a_2}}\phi(x_1)(\lambda^{a_2}x_1, \lambda^{a_1}x_2), \frac{1}{\lambda^{a_1}}\phi(x_2)(\lambda^{a_2}x_1, \lambda^{a_1}x_2), \underline{t}_2\right)$$

may be viewed as an unfolding of $F_\mu(x_1,x_2,\underline{t}_2)$. As such it is induced by the universal unfolding, i.e. there exists $\phi_\lambda \in \Sigma$ and ν such that:

$$G(\lambda) = F_\mu(\phi_\lambda(x_1),\phi_\lambda(x_2),\nu(\lambda))$$

with $\nu(0) = \underline{t}_2$ and $\nu(1) = \underline{t}_1$, which is what we wanted.

Consider now the map between the local rings corresponding to the singularities $(X_\mu(\underline{t}_1),0)$ and $(X_\mu(\underline{t}_2),0)$ induced by ϕ

$$\phi : C[[x_1,x_2]]/(F_\mu(\underline{t}_1)) \to C[[x_1,x_2]]/(F_\mu(\underline{t}_2))$$

and the corresponding map $\overline{\phi}$ of their normalizations

$$\begin{array}{ccc}
C[[t]] & \xrightarrow{\overline{\phi}} & C[[t]] \\
\cup & & \cup \\
C[[t^{a_2} + \text{higher order}, t^{a_1} + \text{higher order}]] & & C[[t^{a_2} + \text{higher order}, t^{a_1} + \text{higher order}]] \\
\| & & \| \\
C[[x_1,x_2]]/(F_\mu(\underline{t}_1)) & \xrightarrow{\phi} & C[[x_1,x_2]]/(F_\mu(\underline{t}_2))
\end{array}$$

We must necessarily have $\overline{\phi}(t) = t\cdot h(t)$, with $h(t)$ a unit in $C[[t]]$. Since $x_1 = t^{a_2}$+ higher order terms, and $x_2 = t^{a_1}$+ higher order terms, it is clear that $\deg \phi \geq 0$.

Q.E.D.

Using this lemma we see that set theoretically $\underline{T}_{\Gamma,\tau} = \underline{S}_\tau/V_\mu$, where $\underline{S}_\tau$ is the open stratum of the flattening stratification of $\underline{H}_\mu$ and V_μ is the kernel of the Kodaira-Spencer map.

We shall prove that $\underline{S}_\tau/V_\mu$ is, locally, an open subset in a weighted projective space.

V_μ is a graded Lie algebra generated as H_μ-module by the elements $\delta_{\underline{\alpha}}$, $\deg \delta_{\underline{\alpha}} = |\alpha|$, see (4.5).

Let V^+ be the sub Lie algebra of all vectorfields of V_μ of degree ≥ 0. V^+ is a finite dimensional solvable C-Lie algebra and $V = [V^+,V^+]$ is nilpotent. Since V^+ generates V_μ as an S_τ-module it is enough to study $\underline{S}_\tau/V^+$.

Notice that the algebraic group $G^+ = \exp V^+$ acts rationally on $\underline{S}_\tau$. The orbits of G^+ are the maximal integral manifolds of V^+, i.e.

the maximal integral manifolds of V_μ.

$G = \exp V$ is a normal subgroup of G^+, $G^+/G = C^*$ and the C^*-action is induced by the Euler vector field $\delta_{\underline{o}}$ of V^+. G acts homogenously (with respect to the C^*-action) on $\underline{S}_\tau$. We shall prove that $\underline{S}_\tau/V = \underline{S}_\tau/G$ is a smooth algebraic variety. The singularities of $\underline{T}_{\Gamma,\tau} = (\underline{S}_\tau/G)/C^*$ are therefore the singularities of the corresponding weighted projective space. We will have to describe, more precisely the open set $\underline{S}_\tau$ and the action of V^+, in terms of the matrix K_0 (see §5) corresponding to the generators $\{\delta_\alpha\}$ of V^+. As before $K_0 = (k_{ij})$ is the matrix of the H_μ-linear map $E:M_1 \to M_2$ with respect to the bases $\{\lambda_i\}_{i=1}^r$ of M_1 and $\{\mu_i\}_{i=1}^r$ of M_2. If $\lambda_i = \underline{x}^{\underline{\alpha}_i}$ then $\delta_{\underline{\alpha}_i} = \sum k_{ij} \frac{\partial}{\partial t_j}$.

Let r_0 be maximal such that $\frac{\mu_j}{x_2} \notin \{\mu_k\}_{k=1}^r$ for $j < r_0$. Since $a_1 < a_2$, $\mu_1,\dots,\mu_{r_0}$ are the monomials closest to the face defined by $(a_1,0)$ and $(0,a_2)$ in R_+^2. Notice that for $i>1$, $\delta_{\underline{\alpha}_i} \in \sum_{j>r_0} H_\mu \cdot \frac{\partial}{\partial t_j}$. This implies that $t_1,\dots,t_{r_0}$ are invariant functions under the action of V.

We shall prove the existence of a filtration $\mathcal{F}^\cdot = \{\mathcal{F}^p M_2\}_{p\in Z}$ on M_2, induced by a filtration of the base $\{\mu_j\}$, and compatible with the degree, and also the dual filtration $\check{\mathcal{F}}^\cdot$ on M_1 defined by $\check{\mathcal{F}}^p M_1 = (\mathcal{F}^{-p}M_2)^\perp = \{\gamma\in M_1, \langle\gamma,\sigma\rangle=0 \text{ for all } \sigma\in\mathcal{F}^{-p}M_2\}$ such that:

$$\text{(P)}\quad \begin{aligned} &\mathcal{F}^{-n}M_2 = M_2,\ \mathcal{F}^n M_2 = \{0\} \text{ if } n \gg 0 \\ &\mathcal{F}^p M_2 \supseteq \mathcal{F}^{p+1}M_2 \\ &\mu_i \in \mathcal{F}^p M_2 \text{ implies } \mu_j \in \mathcal{F}^p M_2 \text{ if } j > i \\ &\lambda_i \in \check{\mathcal{F}}^p M_1 \text{ implies } \lambda_i\mu_j \in \mathcal{F}^p M_2 . \end{aligned}$$

Let us denote by K_0^p the part of the matrix K_0 corresponding to the map $gr_pE:gr_pM_1 \to gr_pM_2$ and by $I_p \subseteq H_\mu$ the radical of the ideal generated by the maximal minors of K_0^p with the convention that $I_p = H_\mu$ if $gr_pM_1 = gr_pM_2 = \{0\}$.

Lemma (6.3). There is a filtration $\mathcal{F}^\cdot$ of M_2 satisfying (P) and the following properties:

(0) $rk_{H_\mu} gr_p M_1 \le rk_{H_\mu} gr_p M_2$ for $p \le 0$.

(1) The elements of K_0^p are polynomials in $t_1,\dots,t_{r_0}$.

(2) $I_p = I_{-p}$.

(3) $I_p \subseteq I_{p-1}$ if $p \le -2$, $I_0 \cap I_{-1} \subseteq I_{-2}$.

(4) $\lambda_i \in \mathcal{F}^{\vee p}$ implies $\lambda_i\lambda_j \in \mathcal{F}^{\vee p+1}$ if $\lambda_i\lambda_j \in \{\lambda_k\}_{k=1,\dots,r}$, where $r = \dim \underline{H}_\mu$, and $j>1$.

(5) $\underline{S}_\tau$ is the open set defined by $I_o \cap I_{-1}$.

Consider first the example $x_1^5+x_2^{11}$, see (§5): We shall be guided by the following facts. Consider the linear matrix $L(\underline{t})$.

$$
\begin{array}{ccccccccc}
t_1 & 2t_2 & 3t_3 & 7t_4 & 8t_5 & 12t_6 & 13t_7 & 18t_8 & 23t_9 \\
 & & & 2t_2 & 3t_3 & 7t_4 & 8t_5 & 13t_7 & 18t_8 \\
 & & & & & 2t_2 & 3t_3 & 8t_5 & 13t_7 \\
 & & & & & t_1 & 2t_2 & 7t_4 & 16t_6 \\
 & & & & & & & 3t_3 & 8t_5 \\
 & & & & & & & 2t_2 & 7t_4 \\
 & & & & & & & & 3t_3 \\
 & & & & & & & & 2t_2 \\
 & & & & & & & & t_1
\end{array}
$$

This matrix has maximal rank iff the "leading" submatrices

$$
(t_1 \quad 2t_2 \quad 3t_3), \quad (2t_2 \quad 3t_3), \quad \begin{pmatrix} 2t_2 & 3t_3 \\ t_1 & 2t_2 \end{pmatrix}, \quad \begin{pmatrix} 3t_3 \\ 2t_2 \end{pmatrix}, \quad \begin{pmatrix} 3t_3 \\ 2t_2 \\ t_1 \end{pmatrix}
$$

have maximal rank. Because of the symmetry we have to consider only the first three of them (up to the "middle matrix"). These matrices have the following property: We get the next one from the one before by, cancelling a column, adding a row or just leaving it unchanged. This means that maximal rank of the "middle" matrix implies maximal rank of the other ones. The "leading" submatrices are "small" enough such that their entries are just the obviously invariant functions $t_1, 2t_2, 3t_3$, with respect to the action of V. We will construct a filtration $\mathfrak{F}^{\cdot}$ on M_2 such that the matrix of the corresponding graded map is given by these "leading" submatrices. Notice that the filtration will not in general be strict . In our example the filtration $\mathfrak{F}^{\cdot}$ gives us the (x_1,x_2)-adic filtration:

$\mathfrak{F}^{-3}M_2$ generated by $x_1x_2^9 \ x_1^2x_2^7 \ x_1^3x_2^5 \ x_1^2x_2^8 \ x_1^3x_2^6 \ x_1^2x_2^9 \ x_1^3x_2^7 \ x_1^3x_2^8 \ x_1^3x_2^9$

$\mathfrak{F}^{-2}M_2$ generated by $x_1^2x_2^8 \ x_1^3x_2^6 \ x_1^2x_2^9 \ x_1^3x_2^7 \ x_1^3x_2^8 \ x_1^3x_2^9$

$\mathfrak{F}^{-1}M_2$ generated by $x_1^2x_2^9 \ x_1^3x_2^7 \ x_1^3x_2^8 \ x_1^3x_2^9$

$\mathfrak{F}^{0}M_2 = \mathfrak{F}^{-1}M_2$

$\mathfrak{F}^{1}M_2$ generated by $x_1^3x_2^8 \ x_1^3x_2^9$

$\mathfrak{F}^{2}M_2 = \mathfrak{F}^{1}M_2$

$\mathfrak{F}^{3}M_2$ generated by $x_1^3x_2^9$

$$K_0^{-3} = \frac{1}{55}(t_1, 2t_2, 3t_3)$$

$$K_0^{-2} = \frac{1}{55}(A, B)$$

$$K_0^{-1} = 0$$

$$K_0^{0} = \frac{1}{55}\begin{pmatrix} A & B \\ t_1 & 2t_2 \end{pmatrix}$$

$$K_0^{1} = 0$$

$$K_0^{2} = \frac{1}{55}\begin{pmatrix} B-\frac{9}{11}t_1A \\ A \end{pmatrix}$$

$$K_0^{3} = \frac{1}{55}\begin{pmatrix} B-\frac{9}{11}t_1A \\ A \\ t_1 \end{pmatrix}$$

$I_{-3} = I_3 = (t_1,t_2,t_3)$, $I_{-2} = I_2 = (A,B)$, $I_{-1} = I_1 = k[t_1,\dots,t_9]$

$I_0 = (2t_2A-t_1B)$

$I_{-3} \supseteq I_{-2} \supseteq I_{-1} \cap I_0$

$\tau = 34$ and $\underline{S}_{34}$ is defined by $2t_2A-t_1B \neq 0$.

Remark: Because of (1) the elements of the matrix K_0^p are invariant functions with respect to the action of V.

Lemma (6.4). Let R be a commutative algebra over a field k, and consider r derivations, $\delta_1,\ldots,\delta_r \in \mathrm{Der}_k(R)$ with the following properties:

(i) $[\delta_i,\delta_j] = 0$ for all i,j

(ii) δ_i is nilpotent, i.e. for all $a \in R$ there is an integer $n(a)$ such that $\delta_i^{n(a)}(a) = 0$

(iii) there are $z_1,\ldots,z_r \in R$ such that for all i,j

- $\delta_i(z_j)$ is invariant with respect to the action of the Lie algebra $L = \sum_{i=1}^{r} k\cdot\delta_i$
- $\det(\delta_i(z_j))$ is invertible in R.

Then $R^L[z_1,\ldots,z_r] = R$ and $\mathrm{Spec}\, R \to \mathrm{Spec}\, R^L$ is a geometric quotient.

We shall postpone the proof of (6.3) and (6.4) until the end of this §.

Now let us study the action of V on $\underline{S}_\tau$. Using (5) and (3) and (1) of (6.3) we can cover $\underline{S}_\tau$ by invariant affine open sets defined by the product of a suitable family of minors of K_0^p, $p \leq 0$.

Let $U = \mathrm{Spec}\, C[\underline{t}]_h$, $h = h_1,\ldots,h_\ell$ be one of these affine open sets, h_i being a minor of K_0^i.

Let $i_0,i_1,\ldots,i_{r_1},i_{r_1+1},\ldots,i_{r_2},\ldots,i_{r_\ell} = r$ define the columns and $1 = j_0,j_1,\ldots,j_{s_1},\ldots,j_{s_\ell}$ the rows of K_0 corresponding to these minors, $r_\ell = s_\ell = \mu-\tau-1$. Denote by δ_j the vectorfields corresponding to λ_j. Let $\delta_k \in V$, and $s_i \leq s \leq s_{i+1}-1$. Because of (6.3) (4) and (4.5) (iii) and (P)

$$[\delta_{j_s},\delta_k] \text{ is in the } C[\underline{t}]_h\text{-module generated by } \delta_{j_{s_{i+1}}},\ldots,\delta_{j_{s_\ell}}.$$

Starting with $\delta_{j_{s_{\ell-1}+1}},\ldots,\delta_{j_{s_\ell}}$ we aply (6.4) ℓ times and get

$$C[\underline{t}]_h^V[t_{i_1},\dots,t_{i_{\mu-\tau-1}}] = C[\underline{t}]_h.$$

We may choose homogeneous invariant functions $d_{j_1},\dots,d_{j_{r-\mu+\tau+1}}$ in $C[\underline{t}]_h$ generating $C[\underline{t}]_h^V$ determined by $d_{j_k} = t_{j_k} \bmod (t_{i_1},\dots,t_{i_{r_\ell}})$.

Then $U/V = \operatorname{Spec} C[d_{j_1},\dots,d_{j_{r-\mu+\tau+1}}]_h$ and $x^{a_1} + x^{a_2} + \sum_k d_{j_k}\cdot\mu_{j_k}$ is the corresponding family.

U/V^+ is then the open set defined by h in the corresponding weighted projective space.

Now it is clear that the quotient of the invariant open sets covering $\underline{S}_\tau$, by V^+, glue to a quasismooth scheme $\underline{T}_{\Gamma,\tau}$. The corresponding families also glue in the étale topology.

Let us return to our example $x_1^5+x_2^{11}$ (see §5). We have $\tau = \tau_{min} = 34$, $\underline{S}_\tau = \operatorname{Spec} C[\underline{t}]_{\{t_2A-t_1B\}}$.

Consider U defined by $t_1A(2t_2A-t_1B) =: h$, $i_0 = 1$, $i_1 = 4$, $r_1 = 1$, $i_2 = 6$, $i_3 = 7$, $r_2 = 3$, $i_4 = 8$, $r_3 = 4$, $i_5 = 9$, $r_4 = 5$. We find

$$C[t]_h^V = C[t_1,t_2,t_3,At_5-Bt_4]_h,$$

and the family:

$$x_1^5+x_2^{11}+t_1x_1x_2^9+t_2x_1^2x_2^7+t_3x_1^3x_2^5+(t_5-\tfrac{B}{A}t_4)x_1^3x_2^6.$$

Now $\underline{T}_{\Gamma,\tau} = \underline{S}_\tau/V^+$ is the open set $D_+(2t_2A-t_1B)$ in $P^3_{(1:2:3:10)} = \operatorname{Proj} C[t_1,t_2,t_3,y]$ with $y = At_5 - Bt_4$.

As above we find universal families

$$x_1^5+ x_2^{11}+ t_1x_1x_2^9+ t_2x_1^2x_2^7+ t_3x_1^3x_2^5+ \tfrac{y}{A}x_1^3x_2^6$$

defined on the open set $U_1: A \neq 0$ in $T_{\Gamma,\tau}$ and

$$x_1^5+ x_2^{11}+ t_1x_1x_2^9+ t_2x_1^2x_2^7+ t_3x_1^3x_2^5- \tfrac{y}{B}x_1^2x_2^8$$

defined on the open set $U_2: B \neq 0$ in $\underline{T}_{\Gamma,\tau}$.

One may check, by direct computation, that these two families do not glue - algebraically - on $U_1 \cap U_2$. However, they obviously glue in the étale topology.

Proof of (6.4). We may assume that $\delta_i(z_j) = \delta_{ij} = \begin{cases} 1 & i=j \\ 0 & i\neq j \end{cases}$. In fact let $Z = (z_{ij})$ be the inverse of the matrix $(\delta_i(z_j))$. The z_{ij} are invariant under the action of L. Now put $\bar{z}_i = \sum_j z_{ji} z_j$, then $\delta_i(\bar{z}_j) = \delta_{ij}$.

Let $R_n = \{y \mid \delta_1^{v_1}, \ldots, \delta_r^{v_r} y = 0 \text{ if } v_1+\ldots+v_r \geq n\}$. $R_1 = R^L$ and $y \in R_n$ implies $\delta_i(y) \in R_{n-1}$.

Assume now $R_{n-1} \subseteq R^L[z_1, \ldots, z_r]$ and let $y \in R_n$. Then

$$\delta_i(y) = \sum_{\substack{\underline{v} \\ v_1+\ldots+v_r \leq n-1}} h^i_{\underline{v}} z_1^{v_1}, \ldots, z_r^{v_r}, \text{ with } h^i_{\underline{v}} \in R^L.$$

$\delta_k \delta_i(y) = \delta_i \delta_k(y)$ implies $v_k h^i_{v_1,\ldots,v_i-1,\ldots,v_r} = v_i h^k_{v_1,\ldots,v_k-1,\ldots,v_r}$ for all $\underline{v}$, with $\sum v_i \leq n$. If $\underline{v}$ is given with $\sum v_i \leq n$ and $v_k > 0$, put $h_{\underline{v}} := \frac{1}{v_k} h^k_{v_1,\ldots,v_k-1,\ldots,v_r}$, and $\bar{y} = \sum_{\underline{v}} h_{\underline{v}} z_1^{v_1}, \ldots, z_r^{v_r}$. Then $y - \bar{y} \in R^L$.

Obviously the invariant functions separate the orbits of L, i.e. $\operatorname{Spec} R \to \operatorname{Spec} R^L$ is a geometric quotient (cf. [D-R]).

Q.E.D.

Proof of (6.3). Let us first describe an effective procedure (due to B. Martin) to order the monomial bases of M_1 and M_2 by degree. Choose integers α, β such that $\alpha a_2 + \beta a_1 = a_1 a_2 + 1$, $0 < \alpha < a_1$ and $0 < \beta < a_2$. Define for every integer k,

$$k*(\alpha,\beta) := (\bar{\alpha}, \bar{\beta})$$

with $0 \leq \bar{\alpha} < a_1$, $0 \leq \bar{\beta} < a_2$ and $\bar{\alpha} \equiv k \cdot \alpha (\operatorname{mod} a_1)$, $\bar{\beta} \equiv k \cdot \beta (\operatorname{mod} a_2)$. Put $Z = \{k \mid 0 \leq k \leq a_1 a_2 - 2a_1 - 2a_2\}$. Then one may check that for $k \in Z$ the quasihomogeneous degree of the vector $k*(\alpha,\beta)$ is given by

$$|k*(\alpha,\beta)| = \begin{cases} k/a_1 a_2 & \text{if } |k*(\alpha,\beta)| < 1 \\ 1 + k/a_1 a_2 & \text{if } |k*(\alpha,\beta)| > 1. \end{cases}$$

Since for any element $x_1^{\alpha_i} x_2^{\beta_i}$ of our base of M_1 (resp. M_2) there is a unique $k_i \in Z$ such that $k_i*(\alpha,\beta) = (\alpha_i, \beta_i)$, we obtain an effective procedure for ordering the basis of M_1 (resp. M_2) by degree. We also get information about the matrix K_0 of $E: M_1 \to M_2$ with respect to the basis $\{\lambda_i\}_{i \leq r}$ of M_1 and $\{\mu_i\}_{i \leq r}$ of M_2. Recall that $K_0 = (k_{ij})$ where $\lambda_i E = \sum k_{ij} \mu_j$.

Let $\lambda_i = x^{\alpha_i}y^{\beta_i}$ correspond to k_i, i.e. $\alpha_i \equiv k_i\alpha \bmod a_1$, $\beta_i \equiv k_i \bmod a_2$, and $\mu_j = x^{\sigma_j}y^{\delta_j}$ correspond to k_j, then $k_{ij} \neq 0$ iff $k_i < k_j$. In the example $x_1^5+x_2^{11}$ we have $\alpha = 1$ and $\beta = 9$

$0*(1,9) = (0,0)$	λ_1	
$1*(1,9) = (1,9)$		μ_1
$2*(1,9) = (2,7)$		μ_2
$3*(1,9) = (3,5)$		μ_3
$4*(1,9) = (4,3)$		
$5*(1,9) = (0,1)$	λ_2	
$6*(1,9) = (1,10)$		
$7*(1,9) = (2,8)$		μ_4
$8*(1,9) = (3,6)$		μ_5
$9*(1,9) = (4,4)$		
$10*(1,9) = (0,2)$	λ_3	
$11*(1,9) = (1,0)$	λ_4	
$12*(1,9) = (2,9)$		μ_6
$13*(1,9) = (3,7)$		μ_7
$14*(1,9) = (4,5)$		
$15*(1,9) = (0,3)$	λ_5	
$16*(1,9) = (1,1)$	λ_6	
$17*(1,9) = (2,10)$		
$18*(1,9) = (3,8)$		μ_8
$19*(1,9) = (4,6)$		
$20*(1,9) = (0,4)$	λ_7	
$21*(1,9) = (1,2)$	λ_8	
$22*(1,9) = (2,0)$	λ_9	
$23*(1,9) = (3,9)$		μ_9

For $k \in Z$ we define its dual $\check{k} \in Z$ by $\check{k} := a_1a_2-2a_1-2a_2-k$. This definition is compatible with the duality on H_μ defined in §5: If $k_i*(\alpha,\beta) = (\alpha_i,\beta_i)$ and $\lambda_i = x_1^{\alpha_i}x_2^{\beta_i}$ then $\check{\lambda}_i = \mu_{r-i+1}$ (cf. §5) corresponds to $\check{k}_i$, i.e. $\check{k}_i*(\alpha,\beta) = (\check{\alpha}_i,\check{\beta}_i)$ and $\mu_{r-i+1} = x_1^{\check{\alpha}_i} x_2^{\check{\beta}_i}$. We will use this description of the monomial basis to construct our filtration.

Put

$$Z_u = \{k_i \mid k_i \in Z,\ \mu_i = x^{k_i\alpha \bmod a_1} y^{k_i\beta \bmod a_2}\}_{i=1,\dots,r}\ ,$$

$$Z_\ell = \{k_i \mid k_i \in Z,\ \lambda_i = x^{k_i\alpha \bmod a_1} y^{k_i\beta \bmod a_2}\}_{i=1,\dots,r}\ ,$$

$$Z_0 = Z \setminus (Z_\ell \cup Z_u).$$

Then $\quad Z_0 = Z_0^{a_1} \cup Z_0^{a_2}$

where $Z_0^{a_1} = \{k \in Z \mid k\alpha \equiv a_1-1 \bmod a_1\}$ and $Z_0^{a_2} = \{k \in Z \mid k\beta \equiv a_2-1 \bmod a_2\}$.

Lemma (6.5).

(1) $k \in Z_0$ iff $\check{k} \in Z_0$

(2) $k \in Z_\ell$ implies $k+a_1 \in Z_\ell$ and either $k-a_1 \in Z_\ell$ or $k-a_1 \in Z_0^{a_2}$

(3) $k \in Z_u$ implies $k-a_1 \in Z_u$ and either $k+a_1 \in Z_u$ or $k+a_1 \in Z_0^{a_2}$

(4) $k \in Z_0^{a_1}$ implies $k \pm a_1 \in Z_0^{a_1}$

(5) $k \in Z_0^{a_2}$ implies $k \pm a_2 \in Z_0^{a_2}$.

Proof: (4) and (5) are trivial. (1) holds because $a_2\alpha + a_1\beta = a_1a_2+1$
To prove (2), let $k \in Z_\ell$, $k*(\alpha,\beta) = (\alpha_1,\alpha_2)$.
Then $0 \leq \alpha_1 < a_1-1$, $0 \leq \alpha_2 < a_2-1$ and $k = a_2\alpha_1 + a_1\alpha_2$.
Now $a_1*(\alpha,\beta) = (0,1)$. Suppose $\alpha_2 = a_2-2$ then $k = a_2\alpha_1 + a_1a_2 - 2a_2$, but $k \leq a_1a_2 - 2a_2 - 2a_1$. Because $\alpha_2 < a_2-2$ we get $(k+a_1)*(\alpha,\beta) = (\alpha_1,\alpha_2+1)$ and $k+a_1 \in Z_\ell$.
If $\alpha_2 > 0$ then $(k-a_1)*(\alpha,\beta) = (\alpha_1,\alpha_2-1)$, i.e. $k-a_1 \in Z_\ell$.
If $\alpha_2 = 0$ then $(k-a_1)*(\alpha,\beta) = (\alpha_1,a_2-1)$, i.e. $k-a_1 \in Z_0^{a_2}$.
(3) is similar. Q.E.D.

Denote by $u(k) = \#\{h \in Z_u \mid h < k\}$ and by $\ell(k) = \#\{h \in Z_\ell \mid h < k\}$. Then $d(k) := u(k) - \ell(k)$ gives the rank of the linear matrix. In fact, let

$d = \max\{d(k) \mid k \leq \frac{1}{2}a_1a_2-a_1-a_2\}$ then the maximal diagonal containing no zeros in the linear matrix, has length $r-d$.

<u>Lemma (6.6)</u>. Let $k \leq \frac{1}{2}a_1a_2-a_1-a_2$ be maximal with the property that $d(k) = d$ then $\check{k}-k \leq a_1+1$.

<u>Proof</u>: Let us suppose $\check{k}-k > a_1+1$, i.e.

$$a_1a_2-2a_1-2a_2-2k > a_1+1$$

$$k < \frac{1}{2}a_1a_2-\frac{3}{2}a_1-a_2-1.$$

We shall show that for any $k < \frac{1}{2}a_1a_2-\frac{3}{2}a_1-a_2-1$ we have the following inequality

$$(*): \# Z_u \cap \{k+1,\ldots,k+a_1\} > \# Z_\ell \cap \{k+1,\ldots,k+a_1\}.$$

If this is true let $h \leq k+a_1$ be maximal such that $h \in Z_u$, and $d(h) = d$. Because of the choice of k we have $h > \frac{1}{2}a_1a_2-a_1-a_2$, i.e. $\check{h} < \frac{1}{2}a_1a_2-a_1-a_2$. Now $d(\check{h})=d(h)-1$ and $\check{h} \in Z_\ell$. This implies $d(\check{h}-1)=d$, i.e. $\check{h}-1 \leq k$, therefore $h > (k+1)^\vee = \check{k}-1$ but $h \leq k+a_1$, i.e. $a_1+1 > \check{k}-k$.

Now we have to prove $(*)$. Since $(\alpha,a_1) = 1$ there exists a permutation $k_0,\ldots,k_{a_1-1}$ of $k+1,\ldots,k+a_1$, such that $k_i*(\alpha,\beta) = (i,c_i)$, $i = 0,\ldots,a_1-1$.

<u>Lemma (6.7)</u>. Suppose $k < \frac{1}{2}a_1a_2-\frac{3}{2}a_1-a_2$ then

(1) $k_i \in Z_\ell$ implies $k_j \in Z_\ell$ for $j < i$

(2) $k_i \in Z_u$ implies $k_j \in Z_u$ for $i < j < a_1-2$

(3) $k_i \in Z_0$ implies $i = a_1-1$ or $i < \frac{1}{2}(a_1-1)$

<u>Proof</u>: (1) Let us assume that $k_j \notin Z_\ell$ for some $j < i$ then $k_j = j\cdot a_2+c_j\cdot a_1-a_1a_2$, $k_i = i\cdot a_2+c_i\cdot a_1$. But this implies $|k_j-k_i|>a_1$ which is a contradiction.

(2) is similar to (1)

(3) if $k_i \in Z_0$ and $i < a_1-1$ then $c_i = a_2-1$, i.e.

$$k_i = i\cdot a_2+(a_2-1)a_1-a_1a_2 = ia_2-a_1.$$

If $i > \frac{1}{2}(a_1-1)$ then $k_i > \frac{1}{2}a_1a_2-\frac{1}{2}a_2-a_1$. But $k_i < \frac{1}{2}a_1a_2-\frac{1}{2}a_1-a_2$, therefore $i < \frac{1}{2}(a_1-1)$. Q.E.D.

Now (*) is an immediate consequence of (6.7). Suppose $Z_\ell \cap \{k+1,\ldots,k+a_1\} = \{k_0,\ldots,k_{t-1}\}$ then $\{k_{t+1},\ldots,t_{a_1-2}\} \subseteq Z_u$ and either $k_t \in Z_u$ or $k_t \in Z_0$. If $k_t \in Z_0$ then $t < \frac{1}{2}(a_1-1)$, i.e.

$$\# Z_\ell \cap \{k+1,\ldots,k+a_1\} = t < a_1-t-2 = \# Z_u \cap \{k+1,\ldots,k+a_1\}.$$

If $k_t \notin Z_0$ then $\# Z_u \cap \{k+1,\ldots,k+a_1\} = a_1-t-1$ and we have to prove that $t < \frac{1}{2}(a_1-1)$ i.e. $k_{[\frac{1}{2}(a_1-1)]} \in Z_u$.

Suppose $k_{[\frac{1}{2}(a_1-1)]} \in Z_\ell$ then $k_{[\frac{1}{2}(a_1-1)]} = a_2[\frac{1}{2}(a_1-1)]+a_1 c_{[\frac{1}{2}(a_1-1)]} > \frac{1}{2}a_1a_2-a_2$. But $k_{[\frac{1}{2}(a_1-1)]} < \frac{1}{2}a_1a_2-a_2-\frac{1}{2}a_1$ gives a contradiction.

Q.E.D.

We are now ready to construct the filtration. We start by constructing a filtration on Z.

Let k be maximal with the property $d(k) = d$ and $k < \frac{1}{2}a_1a_2-a_1-a_2$. By (6.6) we get $\check{k}-k < a_1+1$.

Let
$$H^0 = \{k+1,\ldots,\check{k}-1\}$$
$$H^{-1} = \{\check{k}-a_1,\ldots,k\}$$
$$H^{-i} = \{\check{k}-ia_1,\ldots,\check{k}-(i-1)a_1-1\}\cap Z,\ i\geq 2$$
$$H^i = \check{H}^{-i},\ i>0.$$

Let $\mathcal{F}^{-n}Z = Z$ for $n>>0$ and let $\mathcal{F}^\cdot$ be the filtration induced by the decomposition $Z = \bigcup_{i\in Z} H^i$, such that $\mathcal{F}^pZ = \mathcal{F}^{p-1}\setminus H^{p-1}$.

In our example $x_1^5+x_2^{11}$, $d = 3$, $k = 9$, $\check{k} = 14$

$H^0 = \{10,11,12,13\}$, $H^{-1} = \{9\}$, $H^{-2} = \{4,5,6,7,8\}$, $H^{-3} = \{0,1,2,3\}$, $H^{-4} = \emptyset$.

Notice that it is possible that $H^0 = \emptyset$ or $H^{-1} \subseteq Z_0$ (as in the example). In these cases we find the (x_1,x_2)-adic filtration on M_2. $\mathcal{F}^\cdot$ induces filtrations on Z_u and Z_ℓ:

$$\mathcal{F}^\cdot Z_u = \mathcal{F}^\cdot Z\cap Z_u,\quad \mathcal{F}^\cdot Z_\ell = \mathcal{F}^\cdot Z\cap Z_\ell.$$

Let us denote the induced filtration on M_2 also by $\mathcal{F}^\cdot$ and the induced filtration on M_1 by $\mathcal{F}^{\vee\cdot}$. Notice that $\check{\mathcal{F}}^\cdot M_1 = (\mathcal{F}^{-\cdot} M_2)^\perp$. Notice that $H^0 = \emptyset$ implies $\mathcal{F}^0 = \mathcal{F}^1$ and $H^{-1} \subseteq Z_0$ implies $\mathcal{F}^{-1}Z_u = \mathcal{F}^0 Z_u$ and $\mathcal{F}^1 Z_u = \mathcal{F}^2 Z_u$. Apart from these exceptions, the filtration is always strict in the non-trivial region. To avoid these equalities would just complicate the proof.

$\mathcal{F}^{\cdot}$ has the properties required in (6.3):

(P) is obvious

(0) $gr_p M_1$ is generated by $\{x_1^{\alpha_i} x_2^{\beta_i} \mid (\alpha_i, \beta_i) = k_i * (\alpha, \beta), k_i \in gr_p Z_\ell = H^p \cap Z_\ell\}$

$gr_p M_2$ is genereted by $\{x_1^{\alpha_i} x_2^{\beta_i} \mid (\alpha_i, \beta_i) = k_i * (\alpha, \beta), k_i \in gr_p Z_u = H^p \cap Z_u\}$

By definition $\# H^0 \cap Z_u = \# H^0 Z_\ell$ and $\# H^{-1} \cap Z_u \geq \# H^{-1} Z_\ell$.

By (6.6) $\quad \# H^{-i} \cap Z_u \geq \# H^{-i} \cap Z_\ell$ if $i \geq 2$.

(1) holds because the difference between the minimal and the maximal element in H^p is always smaller or equal to $a_1 - 1$: If an element of K_0^p would depend on t_a for some $a > r_0$ (we may assume its linear part is $x \cdot t_a$, $x \neq 0$) and this element is in the column corresponding to $h \in H^p$ then $h - a_1 \in H^p$.

(2) is a consequence of the duality and the fact that the base change that transforms the matrix K_0 into the symmetric matrix (cf. (5.6)) does not change I_p.

(3) is a consequence of (6.5): In fact $gr_p Z = H^p$, and (6.5) implies

(i) $H^{p-1} \cap Z_u = \{k - a_1, k \in H^p \cap Z_u\} \cup L \quad p \leq -2$

$H^{-2} \cap Z_u = \{k - a_1, k \in (H^{-1} \cup H^0) \cap Z_u\} \cup L$ and L is empty or contains just one element.

(ii) $H^{p-1} \cap Z_\ell = \{k - a_1 \mid k \in H^p \cap Z_\ell\} \setminus T \quad p \leq -2$

$H^{-2} \cap Z_\ell = \{k - a_1 \mid k \in (H^{-1} \cup H^0) \cap Z_\ell\} \setminus T$ and T is empty or contains just one element.

Furthermore $\# H^p \cap Z_u \geq \# H^p \cap Z_\ell$ when $p \leq 0$. Let

$$d_p = \begin{cases} \# H^p \cap Z_\ell, & p \leq -2 \\ \# (H^{-1} \cup H^0) \cap Z_\ell & p = -1 \end{cases}$$

I_{p-1} is the radical of the ideal generated by the d_{p-1}-minors of the matrix K_0^{p-1}, $I_{-1}\cap I_0$ is in the radical of the ideal generated by the d_{-1}-minors of the matrix $K_0^{-1}\oplus K_0^0$.

Let $\lambda_{i_1},\dots,\lambda_{i_{d_{p-1}}}$ generate $gr_{p-1}M_1$, $p < 1$. Suppose I_{p-1} vanishes at a point $\underline{t}\in \underline{S}_\tau$, then the leading forms of

$\lambda_{i_1}E,\dots,\lambda_{i_{d_{p-1}}}E$ with respect to the graduation, i.e. in gr_pM_2, are dependent. Now because of (ii) the leading forms of $x_2\lambda_{i_1}E,\dots,x_2\lambda_{i_{d_{p-1}}}E$ define rows of the matrix K_0^p (if $p<-3$) resp. $K_0^{-1}\oplus K_0^0$ (if $P=-2$). They are also dependent (i).

This implies that the corresponding d_{p-1}-minors of K_0^p resp. $K_0^{-1}\oplus K_0^0$ vanish. But $d_p>d_{p-1}$ implies that I_p resp. $I_{-1}\cap I_0$ also vanishes at $\underline{t}$.

(4) is obvious by definition of $\mathcal{F}^{\bullet}$.

(5) By the choice of k, $d(k) = d$, the linear matrix has maximal rank $r-d$ at a general point, i.e. the matrix K_0 has rank $r-d$ at a general point. The rank drops if the rank of the graded matrices corresponding to H^{-1} and H^0 drop. Because of (2) and (3) $\underline{S}_\tau$ is defined by $I_{-1}\cap I_0$. Q.E.D.

To illustrate the filtration $\mathcal{F}^{\bullet}$ let us consider another example. Let $f = x_1^5 + x_2^{12}$, then $\alpha = 3$, $\beta = 5$

$0*(3,5) = (0,0)$	λ_1	
$1*(3,5) = (3,5)$		μ_1
$2*(3,5) = (1,10)$		μ_2
$3*(3,5) = (4,3)$		
$4*(3,5) = (2,8)$		μ_3
$5*(3,5) = (0,1)$	λ_2	
$6*(3,5) = (3,6)$		μ_4
$7*(3,5) = (1,11)$		
$8*(3,5) = (4,4)$		
$9*(3,5) = (2,9)$		μ_5
$10*(3,5) = (0,2)$	λ_3	
$11*(3,5) = (3,7)$		μ_6
$12*(3,5) = (1,0)$	λ_4	

$13*(3,5) = (4,5)$

$14*(3,5) = (2,10)$ $\qquad\qquad \mu_7$

$15*(3,5) = (0,3)$ $\qquad \lambda_5$

$16*(3,5) = (3,8)$ $\qquad\qquad \mu_8$

$17*(3,5) = (1,1)$ $\qquad \lambda_6$

$18*(3,5) = (4,6)$

$19*(3,5) = (2,11)$

$20*(3,5) = (0,4)$ $\qquad \lambda_7$

$21*(3,5) = (3,9)$ $\qquad\qquad \mu_9$

$22*(3,5) = (1,2)$ $\qquad \lambda_8$

$23*(3,5) = (4,7)$

$24*(3,5) = (2,0)$ $\qquad \lambda_9$

$25*(3,5) = (0,5)$ $\qquad \lambda_{10}$

$26*(3,5) = (3,10)$ $\qquad\qquad \mu_{10}$

$Z = \{0,\ldots,26\}$ $\quad Z_u = \{1,2,4,6,9,11,14,16,21,26\}$

$Z_\ell = \{0,5,10,12,15,17,20,22,24,25\}$

$Z_0^5 = \{3,8,13,18,23\}$

$Z_0^{12} = \{7,19\}$

$d = 3,\ k = 11,\ \check{k} = 15$

$H^0 = \{12,13,14\},\ H^{-1} = \{10,11\},\ H^{-2} = \{5,6,7,8,9\},$

$H^{-3} = \{0,1,2,3,4\}$

$\mathfrak{F}^{-3}Z = Z$

$\mathfrak{F}^{-2}Z = \{k \in Z, k \geq 5\}$

$\mathfrak{F}^{-1}Z = \{k \in Z, k \geq 10\}$

$\mathfrak{F}^{0}Z = \{k \in Z, k \geq 12\}$

$\mathfrak{F}^{1}Z = \{k \in Z, k \geq 15\}$

$\mathfrak{F}^{2}Z = \{k \in Z, k \geq 17\}$

$\mathfrak{F}^{3}Z = \{k \in Z, k \geq 22\}$

$\mathfrak{F}^{4}Z = \emptyset$

the corresponding $gr_p M_1, gr_p M_2$ are generated by

p	$gr_p M_1$	$gr_p M_2$
-3	λ_1	μ_1, μ_2, μ_3
-2	λ_2	μ_4, μ_5
-1	λ_3	μ_6
0	λ_4	μ_7
1	λ_5	μ_8
2	λ_6, λ_7	μ_9
3	$\lambda_8, \lambda_9, \lambda_{10}$	μ_{10}

This is not the (x_1, x_2)-adic filtration on M_2 resp. the induced one on M_1.

The corresponding linear matrix is:

	μ_1	μ_2	μ_3	μ_4	μ_5	μ_6	μ_7	μ_8	μ_9	μ_{10}
λ_1	t_1	$2t_2$	$4t_3$	$6t_4$	$9t_5$	$11t_6$	$14t_7$	$16t_8$	$21t_9$	$26t_{10}$
λ_2				t_1	$4t_3$	$6t_4$	$9t_5$	$11t_6$	$16t_8$	$21t_9$
λ_3						t_1	$4t_3$	$6t_4$	$11t_6$	$16t_8$
λ_4							$2t_2$	$4t_3$	$9t_5$	$14t_7$
λ_5								t_1	$6t_4$	$11t_6$
λ_6									$4t_3$	$9t_5$
λ_7									t_1	$6t_4$
λ_8										$4t_3$
λ_9										$2t_2$
λ_{10}										t_1

$I_{-1} = (2t_2 - \frac{3}{5}t_1^2)$ $\qquad I_0 = (t_1)$

S_τ: $\quad t_1(2t_2 - \frac{3}{5}t_1^2) \neq 0.$

§7. THE MODULI SUITE OF $x_1^5 + x_2^{11}$

We use the notations of §§5 and 6, and the method of constructing the geometric quotient developed there (see Lemma (6.4)).

Consider the versal μ-constant deformation

$$F_\mu(\underline{t}) = x_1^5 + x_2^{11} + t_1x_1x_2^9 + t_2x_1^2x_2^7 + t_3x_1^3x_2^5 + t_4x_1^2x_2^8 + t_5x_1^3x_2^6 + t_6x_1^2x_2^9 + t_7x_1^3x_2^7 + t_8x_1^3x_2^8 + t_9x_1^3x_2^9$$

and it's base space $\underline{H}_\mu = \operatorname{Spec} C[\underline{t}]$.

In §5 we computed the reduced flattening stratification $\{\underline{S}_\tau\}_{34\leq\tau\leq 40}$ of $\underline{H}_\mu$.

$$\underline{S}_{34} = D(t_1B-2t_2A) = D(A(t_1B-2t_2A),\ B(t_1B-2t_2A))$$

$$\underline{S}_{35} = V(t_1B-2t_2A) \cap D(A,B,I)$$

$$\underline{S}_{36} = V(A,B,t_1(11D-9t_1C)) \cap D(C^2-t_1E,\ D^2-(\tfrac{9}{11})^2t_1^3E)$$

$$\underline{S}_{37} = V(A,B,11D-9t_1C,\ C^2-t_1E) \cap D(t_1,E)$$

$$\underline{S}_{38} = V(t_1,t_2,t_3,t_4,t_5,t_7) \cap D(t_6,t_8)$$

$$\underline{S}_{39} = V(t_1,\ldots,t_8) \cap D(t_9)$$

$$\underline{S}_{40} = V(t_1,\ldots,t_9)$$

$$A = 2t_2 - \frac{9}{11}t_1^2$$

$$B = 3t_3 - \frac{7}{11}t_1t_2$$

$$C = 7t_4 + \frac{3}{11}t_1t_2^2$$

$$D = 8t_5 - \frac{8}{11}t_1t_4 + \frac{2}{11}t_1^2t_3^2$$

$$E = 13t_7 - \frac{117}{11}t_1t_6 + \frac{3}{11}t_1^2t_3t_5 + \frac{55}{11^2}t_1t_2t_3t_4 + \frac{7}{5\cdot 11^3}t_1^3t_2^2t_3^2$$

$$I = t_1(A+\frac{1}{11}t_1^2)((t_1D-2t_2C)^2 - 2A(A+\frac{1}{11}t_1^2)(t_1t_7-2t_2t_6))$$

We have also computed vector fields $\{\delta_i\}_{i=1}^9$ of degree ≥ 0 generating the kernel of the Kodaira-Spencer map V_μ as a $C[\underline{t}]$-module:

$$\delta_1 = t_1\frac{\partial}{\partial t_1} + 2t_2\frac{\partial}{\partial t_2} + 3t_3\frac{\partial}{\partial t_3} + 7t_4\frac{\partial}{\partial t_4} + 8t_5\frac{\partial}{\partial t_5} + 12t_6\frac{\partial}{\partial t_6} + 13t_7\frac{\partial}{\partial t_7} + 18t_8\frac{\partial}{\partial t_8} + 23t_9\frac{\partial}{\partial t_9}$$

$$\delta_2 = A\frac{\partial}{\partial t_4} + B\frac{\partial}{\partial t_5} + C\frac{\partial}{\partial t_6} + D\frac{\partial}{\partial t_7} + E\frac{\partial}{\partial t_8} + 18t_8\frac{\partial}{\partial t_9}$$

$$\delta_3 = 2t_2\frac{\partial}{\partial t_6} + (3t_3 + t_1t_2)\frac{\partial}{\partial t_7} + D\frac{\partial}{\partial t_8} + 13t_7\frac{\partial}{\partial t_9}$$

$$\delta_4 = t_1\frac{\partial}{\partial t_6} + 2t_2\frac{\partial}{\partial t_7} + C\frac{\partial}{\partial t_8} + 12t_6\frac{\partial}{\partial t_9}$$

$$\delta_5 = B\frac{\partial}{\partial t_8} + 8t_5\frac{\partial}{\partial t_9}$$

$$\delta_6 = A\frac{\partial}{\partial t_8} + 7t_4\frac{\partial}{\partial t_9}$$

$$\delta_7 = 3t_3\frac{\partial}{\partial t_9}$$

$$\delta_8 = 2t_2\frac{\partial}{\partial t_9}$$

$$\delta_9 = t_1\frac{\partial}{\partial t_9}$$

The aim of this appendix is to analyze the structure of the set $\underline{S}_\tau/V_\mu$. We shall see that there exists a geometric quotient (see [D-R]), $\underline{S}_\tau/V_\mu$ iff $\tau \neq 35$. If $\tau = 35$, $\underline{S}_\tau/V_\mu$ has a natural structure of an algebraic space, but not of a scheme.

As in §6 let V^+ be the (finite dimensional) sub Lie algebra of all vector fields of V_μ of degree > 0 and $V = [V^+,V^+]$. Notice that the polynomials defining the open subset $\underline{S}_\tau$ in its closure $\bar{\underline{S}}_\tau$ are invariant functions on $\bar{\underline{S}}_\tau$ under the action of V and quasihomogeneous with respect to the C^*-action defined by δ_1. Let H be any of the polynomials above defining an affine open (and invariant) subset of $\underline{S}_\tau$. If $H \neq I$ (in the case $\tau = 35$) it is not difficult to see that one can use Lemma (6.4) to compute generators of the ring $\Gamma(\underline{S}_\tau, D(H))^V$ of invariant functions with respect to the action of V on $D(H) \cap \underline{S}_\tau$.

As in the proof of (6.1) we are always able to choose a sequence of sub Lie algebras $V = V_1 \supset \ldots \supset V_r = \{0\}$ such that on $D(H) \cap \underline{S}_\tau$, $[V,V_i] \subseteq V_{i+1}$, and such that V_i/V_{i+1} satisfies the conditions of Lemma (6.4), with respect to its action on $\Gamma(\underline{S}_\tau, D(H))^{V_{i+1}}$. In particular we therefore know that the invariant functions on $D(H) \cap \underline{S}_\tau$ separate the orbits of the action on V, i.e. there exists a geometric quotient, $D(H) \cap \underline{S}_\tau \to D(H) \cap \underline{S}_\tau/V = \operatorname{Spec} \Gamma(\underline{S}_\tau, D(H))^V$. Consequently the morphism

$$D(H) \cap \underline{S}_\tau \to D(H) \cap \underline{S}_\tau / V^+ = \mathrm{Proj}\ \Gamma(\underline{S}_\tau, D(H))^V$$

is a geometric quotient.

For $\tau \neq 35$ it is possible to glue together these quotients to obtain a geometric quotient $\underline{S}_\tau \to \underline{S}_\tau/V^+ =: \underline{T}_\tau$. On $D(H) \cap \underline{S}_\tau/V^+$ we are also able to compute, explicitely, the universal family.

If $\tau = 35$ we shall see that although the ring of invariant functions $\Gamma(\underline{S}_{35}, D(I))^V$ is finitely generated, the corresponding quotient is not geometric, since the invariant functions do not separate the orbits.

Replacing $\underline{S}_{35} \cap D(I)$ by a suitable étale neighbourhood and lifting the action of V^+, we do obtain a geometric quotient, and therefore a structure of algebraic space on the set $\underline{T}_{35} := \underline{S}_{35}/V^+$. Notice that, because of Lemma (6.2) and (3.24) $\underline{T}_\tau$ is equal to the μ-constant stratum of $\underline{N}_\tau$ which is therefore (with the notations of (4.3)) the μ-constant stratum of the reduced moduli suite of $x_1^5 + x_2^{11}$.

Finally $\bigcup_\tau \underline{T}_\tau$ is the coarse moduli space of all plane curve singularities with the semigroup $\Gamma = \langle 5,11 \rangle$.

<u>$\tau = 34$</u>

$\underline{T}_{34} = \underline{S}_{34}/V^+ = D(2t_2A - t_1B) \subseteq \mathrm{Proj}\ \mathbb{C}[t_1, t_2, t_3, y] = \mathbb{P}^3_{(1:2:3:10)}$ with $y = At_5 - Bt_4$ and the universal families

$$x_1^5 + x_2^{11} + t_1x_1x_2^9 + t_2x_1^2x_2^7 + t_3x_1^3x_2^5 + \frac{y}{A}x_1^3x_2^6$$
$$x_1^5 + x_2^{11} + t_1x_1x_2^9 + t_2x_1^2x_2^7 + t_3x_1^3x_2^5 - \frac{y}{B}x_1^2x_2^8$$

defined on $D(A) \cap \underline{T}_{34}$, resp. $D(B) \cap \underline{T}_{34}$.

<u>$\tau = 35$</u>

$\underline{S}_{35} \cap D(A)/V^+ = V(2t_2A - t_1B) \cap D(A) \subseteq \mathrm{Proj}\ \mathbb{C}[t_1, t_2, t_3, x, y] = \mathbb{P}^4_{(1:2:3:8:13)}$

with $x = t_5 - \frac{B}{A}t_4$

$$y = t_7 - \frac{B}{A}t_6 + \frac{1}{A^2}t_4(BC - AD + t_4(\tfrac{1}{2}B - \tfrac{4}{11}t_1A))$$

and the universal family

$$x_1^5 + x^{11} + t_1x_1x^9 + t\ x^2x^7 + t_3x_1^3x_2^5 + x\ x_1^3x_2^6 + y\ x_1^3x_2^7$$

$\underline{S}_{35} \cap D(B)/V^{+} = V(2t_2A-t_1B) \cap D(A) \subseteq \operatorname{Proj} C[t_1,t_2,t_3,u,v] = P^4_{(1:2:3:7:12)}$

with $u = t_4 - \frac{A}{B}\, t_5$

$$v = t_6 + \frac{A}{B}\, t_7 - \frac{1}{B^3}\, t_5(ABD-B^2C + t_5(\tfrac{4}{11}t_1A^2 - \tfrac{1}{2}AB))$$

and the universal family

$$x_1^5 + x_2^{11} + t_1x_1x_2^9 + t_2x_1^2x_2^7 + t_3x_1^3x_2^5 + u\, x_1^2x_2^8 + v\, x_1^2x_2^9$$

Notice that

$$Ax = -Bu,\quad -AB^2y + A(\tfrac{15}{2}B + \tfrac{4}{11}t_1A)x^2 - \tfrac{8}{11}t_3^2(3t_2 - t_1^2)ABx = B^3v.$$

$S_{35} \cap D(I)/V^{+}$ does not exist as a geometric quotient:

$$\Gamma(\underline{S}_{35}, D(I)))^V = C[t_1,t_2,t_3,,I_0,I_1]_I/(2t_2A - t_1B)$$

$$I_0 = t_1t_5 - 2t_2t_4$$

$$I_1 = t_1A\, y = A(t_1t_7 - 2t_2t_6) + t_4(2t_2C - t_1D + t_4(t_2 - \tfrac{4}{11}t_1^2))$$

Notice that since $t_1B = 2t_2A$

$$t_1I = (-2t_1^2(A + \tfrac{1}{11}t_1^2)I_1 + (8t_1I_0 - \tfrac{2}{99}t_2^2(4t_2 - t_1^2)^2(3t_2 - t_1^2))^2)\cdot(A + \tfrac{1}{11}t_1^2)$$

and use the fact that

$$\Gamma(\underline{S}_{35}, D(I\cdot A))^V = C[t_1,t_2,t_3,I_0,I_1]_{A\cdot I}/(2t_2A - t_1B)$$

Now, obviously, the orbit (i.e. the maximal integral manifold) of V through a point $\underline{t} \in \underline{S}_{35} \cap V(A,B)$ is $\{(t_1,\ldots,t_5)\}\times C^4$. But since $A = B = 0$

$$I_1(\underline{t}) = \frac{143}{22}\, t_1^2t_4^2 - 8t_1t_4t_5 + \frac{5\cdot 3^2\cdot 7^2}{2^2\cdot 11^6}\, t_1^9t_4$$

$$I_0(\underline{t}) = t_1t_5 - \frac{9}{11}\, t_1^2t_4^2$$

In general there are two different points

$$\underline{t}^{(1)} = (t_1,t_2,t_3,t_4^{(1)},t_5^{(1)},0\ldots0)$$

$$\underline{t}^{(2)} = (t_1,t_2,t_3,t_4^{(2)},t_5^{(2)},0\ldots0)$$

such that

$$I_0(\underline{t}^{(1)}) = I_0(\underline{t}^{(2)})$$

$$I_1(\underline{t}^{(1)}) = I_1(\underline{t}^{(2)}) \text{ , i.e.}$$

the invariant functions do not separate the orbits.

The fibres of the map $\underline{S}_{35} \cap D(I) \to \text{Spec } \Gamma(\underline{S}_{35}, D(I))^V$ at points with $A = 0$, consist in general of two orbits. Therefore the geometric quotient does not exist.

Now consider the étale neighbourhood $\underline{E}$ of $\underline{S}_{35} \cap D(I)$ defined by

$$0 = A \cdot z^2 + 2(t_1 D - 2t_2 C)z + 2(A + \frac{1}{11} t_1^2)(t_1 t_7 - 2t_2 t_6)$$

The action of V can be lifted to $\underline{E}$ by

$$\delta_2(z) := -(A + \frac{1}{11} t_1^2)$$

$$\delta_i(z) := 0 \qquad i > 2$$

It is not difficult to see that

$$\Gamma(\underline{E})^V = \mathbb{C}[t_1, t_2, t_3, , I_0, \bar{I}_1]_{\bar{I}_1 \cdot t_1 (A + \frac{1}{11} t_1^2)} / (t_1 B - 2t_2 A)$$

with $\bar{I}_1 = Az + t_1 D - 2t_2 C$

One may also check that the invariant functions are separating the orbits.

The universal family on $\text{Proj } \Gamma(\underline{E})^V$ is

$$x_1^5 + x_2^{11} + t_1 x_1 x_2^9 + t_2 x_1^2 x_2^7 + t_3 x_1^3 x_2^5 + (t_4 + \frac{A\ z}{A + \frac{1}{11} t_1^2}) x_1^2 x_2^8 + (t_5 + \frac{2t_2 Az}{t_1 (A + \frac{1}{11} t_1^2)}) x_1^3 x_2^6$$

<u>$\tau = 36$</u>

$$\underline{S}_{36} \cap D(C^2 - t_1 E)/V^+ = V(t_1(11D - 9t_1 C)) \cap D(C^2 - t_1 E) \subseteq \text{Proj } \mathbb{C}[t_1, t_4, t_5, x]$$

$$= \mathbb{P}^3_{(1:7:8:13)}$$

with $x = t_7 - \dfrac{CD - \frac{9}{11} t_1^2 E}{C^2 - t_1 E} t_6$

(notice that because of $A = B = 0$, i.e. $t_2 = \frac{9}{22}t_1^2$, $t_3 = \frac{21}{2\cdot 11^2}t_1^3$ we may omit t_2, t_3).

The universal family is:

$$x_1^5 + x_2^{11} + t_1x_1x_2^9 + \frac{9}{22}t_1^2x_1^2x_2^7 + \frac{21}{242}t_1^3x_1^3x_2^5 + t_4x_1^2x_2^8 + t_5x_1^3x_2^6 + xx_1^3x_2^7$$

$$\underline{S}_{36} \cap D(D^2-(\tfrac{9}{11})^2t_1^3E)/V^+ = V(t_1(11D-9t_1C)) \cap D(D^2-(\tfrac{9}{11})^2t_1^3E) \subseteq \mathrm{Proj}\ C[t_1,t_4,t_5,u]$$

$$= P^3_{(1:7:8:12)}$$

with

$$u = t_6 = \frac{CD-\frac{9}{11}t_1^2E}{D^2-(\frac{9}{11})^2t_1^3E}t_7$$

The universal family is:

$$x_1^5 + x_2^{11} + t_1x_1x_2^9 + \frac{9}{22}t_1^2x_1^2x_2^7 + \frac{21}{242}t_1^3x_1^3x_2^5 + t_4x_1^2x_2^8 + t_5x_1^3x_2^6 + ux_1^2x_2^9.$$

Notice that $(CD-\frac{9}{11}t_1^2E)x = -(D^2-(\frac{9}{11})^2t_1^3E)u =: w$, i.e.

$$\underline{T}_{36} := \underline{S}_{36}/V^+ = V(t_1(11D-9t_1C)) \cap D(C^2-t_1E, D^2-(\tfrac{9}{11})^2t_1^3E)$$

$$\subseteq \mathrm{Proj}\ C[t_1,t_4,t_5,w] = P^3_{(1:7:8:28)}$$

$\underline{\tau = 37}$

$$\underline{S}_{37} \cap D(t_1)/V^+ = V(C^2-t_1E) \cap D(t_1) \subseteq \mathrm{Proj}\ C[t_1,t_4,x,y] = P^3(1:7:13:18)$$

with

$$x = t_7 - \frac{9}{11}t_1t_6$$

$$y = t_8 - \frac{C}{t_1}t_6$$

(Notice that $E = 13x - \frac{3.109}{16.11^2}t_1^6(C - \frac{21^2}{10.11^4}t_1^7)$.)

The universal family is:

$$x_1^5 + x_2^{11} + t_1x_1x_2^9 + 1\frac{9t_1^2}{22}x_1^2x_2^7 + \frac{21t_1^3}{242}x_1^3x_2^5 + t_4x_1^2x_2^8$$

$$+ (\frac{71t_1}{88}t_4 + \frac{5.21^2}{32.11^4}t_1^8)x_1^3x_2^6 + xx_1^3x_2^7 + yx_1^3x_2^8$$

$$\underline{S}_{37} \cap D(E)/V^+ = V(C^2-t_1, E) \cap D(E) \subseteq \mathrm{Proj}[t_1,t_4,u,v] = P^3_{(1:7:12:13)}$$

with

$$u = t_6 - \frac{C}{E} t_8$$

$$v = t_7 - \frac{9}{11} t_1 \frac{C}{E} t_8$$

The universal family is:

$$x_1^5 + x_2^{11} + t_1 x_1 x_2^9 + \frac{9}{22} t_1^2 x_1^2 x_2^7 + \frac{21}{242} t_1^3 x_1^3 x_2^5 + t_4 x_1^2 x_2^8$$
$$+ \left(\frac{71}{88} t_1 t_4 + \frac{5.21^2}{32.11^4} t_1^8\right) x_1^3 x_2^6 + u x_1^2 x_2^9 + v x_1^3 x_2^7$$

Notice that

$$v - \frac{9}{11} t_1 u = x$$

$$-Cu = t_1 y =: w, \quad \text{i.e.}$$

$$\underline{T}_{37} = \underline{S}_{37}/V^+ = V(C^2 - t_1 E) \cap D(t_1, E) \subseteq \operatorname{Proj} C[t_1, t_4, x, w] = P^3_{(1:7:13:19)}$$

$\tau = 38$

$$\underline{T}_{38} = \underline{S}_{38}/V^+ = \operatorname{Proj} C[t_6, t_8] = P_{(12:18)} \simeq P^1$$

The universal family is:

$$x_1^5 + x_2^{11} + t_6 x_1^2 x_2^9 + t_8 x_1^3 x_2^8$$

$\tau = 39$

$\underline{T}_{39} = \underline{S}_{39}/V^+$ is a point with corresponding singularity $x_1^5 + x_2^{11} + x_1^3 x_2^9$.

$\tau = 40$

$\underline{T}_{40} = \underline{S}_{40}$ is a point with corresponding singularity $x_1^5 + x_2^{11}$.

§8 APPENDIX: (BY B. MARTIN AND G. PFISTER) AN ALGORITHM TO COMPUTE THE KERNEL OF THE KODAIRA-SPENCER MAP FOR AN IRREDUCIBLE PLANE CURVE SINGULARITY

We keep the notation of §5 and §6 and give an algorithm for computing the matrix K_0 defined by $E\lambda_i = \sum h_{ij}\mu_j$ in the case $(a_1,a_2) = 1$ (cf. §5).

To construct the monomial bases $\{\lambda_i\}$ of M_1 and $\{\mu_i\}$ of M_2, we apply the method above using a primitive vector (cf. proof of (6.3)).

<u>Part (i) of the algorithm</u>:

In this part of the algorithm we compute a string S related to $k*(\alpha,\beta)$:

$$S(k) = \begin{cases} \ell & \text{if } k \in Z_\ell \\ * & \text{if } k \in Z_0 \\ u & \text{if } k \in Z_u \end{cases}$$

```
Compute α and β such that a₂α≡1 mod a₁, a₁β≡1 mod a₂;
j:=1; i:=e:=e':=∅; S(1):='ℓ'; ℓ(1):=∅; d:=a₁a₂
FOR k:= 1 TO d-2(a₁+a₂) DO
e:=e+α mod a₁; e':=e'+β mod a₂; e(k):=e; e'(k):=e';
{e(k),e'(k) is needed only in part (iii) of the algorithm}
   IF (e=a₁-1) OR (e'=a₂-1)  THEN  S(k+1):='*'
   ELSIF  (ea₂+e'a₁)<d  THEN  i:=i+1; ℓ(i):=k; S(k+1):='ℓ'
   {characterizes monomials of M₁, ℓ(i) is the degree of λᵢ}
   ELSE  j:=j+1; u(j):=k; S(k+1):='u'
   {characterizes monomials of M₂, u(j)+d is the degree of μⱼ}
   END
END
```

Example: $f = x_1^5+x_2^{14}$, $\alpha = 4$, $\beta = 3$

S = 'ℓ*uuuℓ*uu*ℓ*uuℓℓ*uuℓℓ*uuℓℓ*uuℓℓ*u*ℓℓ*uℓℓℓ*u'

Notice that $S(u) = 'u'$ iff $S(N+1-u) = '\ell'$, N:=length of the string $= a_1a_2-2a_1-2a_2+1$ (duality), i.e. it is sufficient to compute only half the string.

Now a monomial $t_{g(1)}\cdots t_{g(s)}$ occurs in the element h_{ij} of the matrix K_0 iff

(i) $\lambda_i\mu_{g(1)}\cdots\mu_{g(s)} = \mu_j\cdot x^{ma_1}y^{na_2}$ for some m,n such that $m+n=s-1$.

(ii) $\lambda_i\mu_{g(1)}\cdots\mu_{g(t)}/x^{ma_1}y^{na_2}$ is not in the monomial base for $t=1,\ldots,s-1$, and m,n such that $m+n=t-1$.

This is equivalent (see the construction of the string above and the method used in 6.2) to

(i') $\ell(i)+u(g(1))+\ldots+u(g(s)) = u(j)$

(ii') $S(\ell(i)+u(g(1))+\ldots+u(g(t)))\neq 'u'$ for $t = 1,\ldots,s-1$.

Using (i') and (ii') we can compute the monomials of h_{ij}.

<u>Part (ii) of the algorithm</u>: Computing the monomials of h_{ij}.

```
PROCEDURE bracket (r,s);
BEGIN b:=s-r; F:=F+'(';
REPEAT g:=max{i:u(i)<b};
WHILE g>∅ DO b:=u(g)-1;
   IF u(g) = s-r THEN F:=F+'t_g+'
   ELSE IF S(1+r+u(g))≠'u' THEN  F:=F+'t_g'; bracket (r+u(g),s) END;
   g:=max{i:u(i)<b}
END;

IF (last character of F≠'+')  THEN  (delete last two char) END;
UNTIL b = ∅;
Replace last character of F by ')+'
END;
BEGIN u(∅):=∅; F:='';
      bracket (ℓ(i),u(j));
      Delete last character of F
END.
```

Example: $F = x_1^5+x_2^{14}$, $i = 2$, $j = 9$

$F = '(t_7+t_3(t_4(t_1)+t_1(t_4+t_3(t_2)+t_2(t_3+t_1(t_1)))))'$

to finish this part of the algorithm we have to "solve the brackets. In our example we get:

$$F = 't_7+t_3t_4t_1+t_3t_1t_4+t_3t_1t_3t_2+t_3t_1t_2t_3+t_3t_1t_2t_1t_1'$$

Now we have to compute the coefficients of the corresponding monomials. If c is the coefficient of the monomial $t_{g(1)}\cdot\ldots\cdot t_{g(s)}$ occuring in F, then $c = c_1\cdot\ldots\cdot c_s$ and $c_1 = 1-\deg \mu_{g(1)}/d$, c_i is one of the exponents of $\mu_g(i)$ divided by a_1 or a_2 depending on whether $\frac{\partial F_\mu}{\partial x_1}$ resp. $\frac{\partial F_\mu}{\partial x_2}$ was involved in the reduction modulo $(\frac{\partial F_\mu}{\partial x_1}, \frac{\partial F_\mu}{\partial x_2})$ in this step.

Part (iii) of the algorithm: Computing a coefficient

```
c:=(-1)^s u(g(1))/d; n:=ℓ(i);
FOR k:=2 TO s DO
e:=e(n)+e(u(g(k-1))); e':=e'(n)+e'(u(g(k-1))); n:=n+g(k-1);
IF e>a_1-1 THEN c:=c·e(u(g(k)))/a_1  END;
IF e'>a_2-1 THEN c:=c·e'(u(g(k)))/a_2  END;
END
```

Example: $f = x_1^5+x_2^{14}$, $i = 2$, $j = 9$, $t_3t_1t_2t_1t_1$

$$c = -\frac{4}{70}\cdot\frac{6}{14}\cdot\frac{2}{5}\cdot\frac{6}{14}\cdot\frac{3}{5} = -\frac{108}{42875}$$

Part (iv) of the algorithm:

Finally we have to order the variables in the monomials, to order the monomials in F lexicographically and to identify monomials of the same type by adding their coefficients.

Example: $f = x_1^5+x_2^{14}$

$$h_{2,9} = -\frac{13}{70}t_7-\frac{39}{1225}t_4t_3t_1+\frac{99}{8575}t_3^2t_2t_1-\frac{108}{42875}t_3t_2t_1^3$$

Remark: The algorithm can be modified for more than two variables. We can also omit the restriction that a_1 and a_2 be relatively prime.

Bibliography:

[An] André, M.: Méthode simpliciale en algèbre homologique et algèbre commutative. Lecture Notes in Mathematics, Springer Verlag, No. 32 (1976).

[A] Arnold, V.-I. & Varchenko, A.N. & Hussein-Zade, S.M.: Singulatities of Differentiable maps. I, II. Moscow 1982, 1984.

[Ar] Artin, M.: Algebraic approximation of structures over complete local rings. Publ. Math. IHES 36 (1969), pp. 23-58.

[Br] Briançon, J., Granger, M. et Maisonobe, P.: Le nombre de modules du germe de courbe plane $x^a + y^b = 0$. Manuscript, Institut de mathématiques et sciences physiques, Université de Nice, juin 1986.

[Del] Delorme, C.: Sur les modules des singularités des courbes planes. Bull. Soc. Math. de France 106 (4) (1978) pp. 417-446.

[D-G] Demazure, M. & Gabriel, P.: Groupes Algébriques, Tome 1. Masson & Cie. Paris. North Holland, Amsterdam.

[D-R] Dixmier, J., Raynaud, M.: Sur le quotient d'une variété algébrique par un group algebrique. L. Nachbin, Mathematical analysis and applications (Essays dedicated to Laurent Schwartz) Academic Press (1981).

[E] Elkik, R.: Solutions d'équations à coefficients dans un anneau henselien. Ann. Scient. Ecole Normale Supérieure Paris, 4e série, t.b. (1976) pp. 553-604

[Eb] Ebey, S.: The classification of singular points of algebraic curves. Trans Amer. Math. Soc. (1965) pp. 454-471.

[G-H] Greuel, G.-M. & Hamm, H.: Invarianten quasihomogener vollständiger Durchschnitte. Inventiones Math. Vol 49 (1978) pp. 67-86.

[G-L] Greuel, G.-M. & Looijenga, E.: The dimension of smoothing components. Duke Mathematical Journal Vol 52 No 1 (1985) pp. 263-272.

[GR 1] Grothendieck, A.: Eléments de Géometrie Algébrique, Chap. III.

[GR 2] Grothendieck, A.: Technique de descente et théorèmes d'existence en géometrie algébrique, I. Séminaire Bourbaki 12e année 1959/60 No 190.

[Kn] Knutson, D.: Algebraic spaces. Lecture Notes in Mathematics Springer-Verlag, No 203, (1971).

[La-Lø] Laudal, O.A. & Lønsted, K.: Deformations of curves. Moduli for hyperelliptic curves. Algebraic Geometry, Tromsø 1977, Lecture Notes in Mathematics, Springer-Verlag No 687.

[La-Pf] Laudal, O.A. & Pfister, G.: The local moduli problem. Applications to the classification of isolated hypersurface singularities. University of Oslo, Preprint Series No 11 (1983).

[L-M-P] Laudal, O.A. & Martin, B. & Pfister, G.: Moduli of plane curve singularities with C^*-action. To appear in the publication series of the Banach Center, Warsaw (Banach semester 1986), Vol 20 (1987).

[La 1] Laudal, O.A.: Formal moduli of algebraic structures. Lecture Notes in Mathematics, Springer-Verlag No 754 (1979).

[La 2] Laudal, O.A.: Matric Massey products and formal moduli I. Algebra, Algebraic Topology and their interactions. Lecture Notes in Mathematics, Springer-Verlag No 1183 (1986) pp. 218-240.

[M] Mumford, David: Lectures on curves on an algebraic surface Annals of Mathematics Studies 59, Princeton (1966).

[Me] Merle, M. Sur l'espace des modules des courbes de semi-groupe donné C.R. Acad. Sci. Paris 284 (1977) pp. 611-614

[Pal] Palamodov, V.P.: Moduli in versal deformations of complex spaces. Lecture Notes in Mathematics, Springer-Verlag No 683 (1978) pp. 74-115.

[R] Rosenlicht, M.: A remark on quotient spaces. Annales Acad. Brasil Cienc. 35 (1963) pp. 487-489.

[S] Saito, K.: On the periods of primitive integrals I. Rims-preprint 412, Kyoto University (1982).

[Sch] Schlessinger, M.: Functors of Artin Rings. Trans. Amer. Math. Soc. Vol 130 (1986) pp. 208-222.

[W] Wahl, J.: Smoothings of normal surface singularities. Topology Vol 20 (1981) pp. 219-246.

[Wash] Washburn, S.: On the moduli of plane curve singularities I. Journal of Algebra 56 (1) (1979) pp. 91-102, & II, loc.cit. 66. (2) (1980), pp. 354-385.

[Z] Zariski, O.: Le problème des modules pour les branches planes Pub. du Centre de Mathématiques de l'Ecole Polytechnique, Paris (1976).

Subject index

LECTURE NOTES IN MATHEMATICS
Edited by A. Dold and B. Eckmann

Some general remarks on the publication of monographs and seminars

In what follows all references to monographs, are applicable also to multiauthorship volumes such as seminar notes.

1. Lecture Notes aim to report new developments - quickly, informally, and at a high level. Monograph manuscripts should be reasonably self-contained and rounded off. Thus they may, and often will, present not only results of the author but also related work by other people. Furthermore, the manuscripts should provide sufficient motivation, examples and applications. This clearly distinguishes Lecture Notes manuscripts from journal articles which normally are very concise. Articles intended for a journal but too long to be accepted by most journals, usually do not have this "lecture notes" character. For similar reasons it is unusual for Ph.D. theses to be accepted for the Lecture Notes series.

   Experience has shown that English language manuscripts achieve a much wider distribution.

2. Manuscripts or plans for Lecture Notes volumes should be submitted either to one of the series editors or to Springer-Verlag, Heidelberg. These proposals are then refereed. A final decision concerning publication can only be made on the basis of the complete manuscripts, but a preliminary decision can usually be based on partial information: a fairly detailed outline describing the planned contents of each chapter, and an indication of the estimated length, a bibliography, and one or two sample chapters - or a first draft of the manuscript. The editors will try to make the preliminary decision as definite as they can on the basis of the available information.

3. Lecture Notes are printed by photo-offset from typed copy delivered in camera-ready form by the authors. Springer-Verlag provides technical instructions for the preparation of manuscripts, and will also, on request, supply special staionery on which the prescribed typing area is outlined. Careful preparation of the manuscripts will help keep production time short and ensure satisfactory appearance of the finished book. Running titles are not required; if however they are considered necessary, they should be uniform in appearance. We generally advise authors not to start having their final manuscripts specially tpyed beforehand. For professionally typed manuscripts, prepared on the special stationery according to our instructions, Springer-Verlag will, if necessary, contribute towards the typing costs at a fixed rate.

   The actual production of a Lecture Notes volume takes 6-8 weeks.

.../...

4. Final manuscripts should contain at least 100 pages of mathematical text and should include

   - a table of contents
   - an informative introduction, perhaps with some historical remarks. It should be accessible to a reader not particularly familiar with the topic treated.
   - subject index; this is almost always genuinely helpful for the reader.

5. Authors receive a total of 50 free copies of their volume, but no royalties. They are entitled to purchase further copies of their book for their personal use at a discount of 33 1/3 %, other Springer mathematics books at a discount of 20 % directly from Springer-Verlag.

   Commitment to publish is made by letter of intent rather than by signing a formal contract. Springer-Verlag secures the copyright for each volume.